Acknowledgments

Research for this book was supported by generous grants from the Dibner History of Science Program at the Huntington Library, Villa I Tatti, the Gladys Krieble Delmas Foundation, the Fonds voor Wetenschappelijk Onderzoek in conjunction with the De Wulf-Mansion Centre, the History of Science Department at the University of Oklahoma, and Oakland University. I am extremely grateful for their support.

Preliminary versions of several chapters were presented at annual meetings of the Renaissance Society of America and the History of Science Society, at a seminar at the Department of History and Philosophy of Science at Indiana University, and at McGill University. I thank all of those who posed questions at those presentations. I received many comments and criticisms for earlier versions and drafts of chapters. For these suggestions I owe gratitude to Matthew Klemm, Christoph Lüthy, Gideon Manning, Bill Newman, Katharine Park, H. Darrel Rutkin, Andrew Sparling, Edith Sylla, and Gideon Yaffe.

Although the research for this book was conducted at numerous libraries, I spent much time at a few. I thank the staffs of the Biblioteca Ambrosiana, the Biblioteca Marciana, the Biblioteca Nazionale Centrale Vittorio Emanuele II, the Biblioteca Universitaria di Padova, the History of Science Collections at the University of Oklahoma, and the Huntington Library for their assistance and generosity. Fabio Forner helped introduce me to many Italian libraries. I wrote much of this book as a Dibner Fellow at the Huntington Library, where discussions and conversations with Nick Dew, Jan Golinski, Juan Gomez, Roy Laird, Anca Parvulescu, Constantin Parvulescu, Nick Popper, Tiffany Werth, and many others sustained me. I thank Pam Long for encouraging me at the start and at the end of the project. I am grateful for the work of Josh Tong, who greatly improved the final manuscript. From the beginnings of my career to the present, Don Smith has provided me with intellectual inspiration and companionship. Much of my approach to the study of the history of science I owe to the teaching of John Murdoch, my doctoral adviser. It greatly saddens me that John did not live to see this book in print.

Finally, I thank my parents and my brother Russell for their kindness, love, and affection. Without them, I could not have written this book.

Renaissance Meteorology

Introduction

Descriptions of the emergence of modern science frequently depend on identifying breaks between the new natural philosophies of the seventeenth century and scholastic Aristotelian natural philosophy. Seventeenth-century proponents of new natural philosophies, such as Francis Bacon, René Descartes, and Galileo Galilei, criticized Aristotelians for their reliance on texts and neglect of experience, their adherence to metaphysical concepts at the expense of considerations of nature, their demand for syllogistic and certain knowledge, their unwillingness to incorporate new ideas, and their lack of concern for practical matters. Attacking Aristotelians for their alleged excessive reliance on syllogism and abstraction was a trope of humanist authors beginning as early as the 1350s with Francesco Petrarca.[1] And examples of Aristotelians concerning themselves with logic and metaphysics abound, not surprisingly, in commentaries on Aristotle's works on metaphysics and logic.

Humanists like Petrarca, however, did not always have deep knowledge of natural philosophy.[2] Many of the stereotypical notions of scholasticism do not apply to all fields of traditional natural philosophy and do not apply in particular to Renaissance meteorology. A comprehensive examination of Renaissance natural philosophy, which includes meteorology, shows that Aristotelians were more than bookish pedants. Early modern authors of meteorological treatises recognized that understandings of meteorological phenomena should be flexible and limited and that knowledge of the causes of the weather was inherently uncertain. Despite this uncertainty, meteorology was seen as an endeavor that was extremely useful, not just for manual activities but also for the needs of

Ages and the Renaissance, despite differences in context, style, and intent, shared a primary interest in causes and explanations.

The focus on causes was central to natural philosophy and a legacy of the influence of Aristotle's *libri naturales* from the Middle Ages to the middle of the seventeenth century. In Aristotelian natural philosophy, causation is identical to explanation; the Greek word *aition* was originally a legal term that referred to the responsibility for a particular event or crime.[5] Transferred to the realm of natural philosophy, causes were supposed to give an account of natural processes and substances. Causation was conceptually far broader for Aristotle and for nearly all of premodern natural philosophy than for contemporary natural sciences. Aristotle, famously, enumerated four kinds of causes or explanations: material, efficient, formal, and final. Material causes are the constituents of something, efficient causes are what provoke the motion of matter, formal causes are equivalent to the essences or organizing principles of substances, and final causes describe the purposes and ends of substances. It is with these four causes that most of meteorology until the time of Descartes and even beyond was concerned.

During the twelfth and thirteenth centuries, scholars translated the Aristotelian corpus, first from Arabic and then from the original Greek. As universities emerged during the same period, they incorporated these Latin translations into their curricula, where they formed the basis for instruction in logic and natural philosophy in faculties of arts. They remained the basis until the late seventeenth century. Interpreting and commenting on Aristotle's works became a mode not just of instruction but also of doing natural philosophy. Curricula varied from one university to the next, but, by and large, bachelor students were required to follow lectures on *Physics, De generatione et corruptione, De caelo, Meteorology, De anima,* and parts of *Parva naturalia.* This undergraduate instruction laid the foundation for graduate studies, especially in the faculties of medicine and theology.

Although there was much continuity in the Aristotelian tradition from 1200–1650, it was by no means static or uniform. Lecturing or commenting on Aristotle in no way meant agreeing with his view. Natural philosophers found his stances at times not in accordance with experience, reason, or the demands of theology. Moreover, the emphasis on causation did not end with the rejection of Aristotelian thought: his sharpest critics devoted themselves to similar explorations. The field of meteorology was largely devoted to reporting, describing, and explaining extraordinary meteorological phenomena even through much of the eighteenth century, when eventually the ideal of the laboratory and the desires of agricultural operators for scientific prognostics displaced earlier modes of meteorology.[6]

Much of Renaissance and early modern meteorology was produced in university settings. It was not the most prominent field despite being firmly situated in university curricula. In a perhaps apocryphal anecdote, the humanist scholar Francesco Spino wrote in a letter to Pier Vettori that after listening to Simone Porzio, a professor at the University of Pisa during the 1540s and 1550s, inaugurate his lectures with a discourse on the *Meteorology,* students began to clamor for Porzio to lecture on the more controversial third book of the *De anima,* chanting "anima, anima."[7] While the exact meaning of this incident is elusive, it does suggest that these students were more enthusiastic about hearing Porzio's materialist reading of Aristotle's psychology than his meteorology.[8] The students' enthusiasm mirrored institutional norms, as meteorology was in Italy generally part of the duties of extraordinary professors or a subject for professors who lectured on holidays.[9] Nevertheless, Porzio, or his overseers, thought that meteorology was of sufficient relevance that he began his lectures on that topic. But it would be inaccurate to claim that the prominence of meteorology was equal to that of topics found in *De anima, Physics,* or *Metaphysics,* although meteorology was considered a crucial and necessary part of knowledge of the natural world and the university curriculum.

THE PLACE AND SCOPE OF RENAISSANCE METEOROLOGY

Like many fields, meteorology during the Renaissance differed from current conceptions. A sixteenth-century German, Marcus Frytsche, began his textbook on meteorology with a question and response: "What is meteorology?" he asked. The answer was, "It is the part of physics that is concerned with what comes to be in the regions of the air or in the belly of the earth."[10] While the contemporary world is apt to connect the field of meteorology with prediction, Frytsche's definition demonstrates that meteorology was the part of physics, or natural philosophy, concerned with what occurs in a particular part of the universe, namely the sublunary region. Most approaches to early modern meteorology were concerned not with prognostication but with explanation and causation, the dual purposes of all of physics or natural philosophy. This distinction is not arbitrary but rather reflective of divisions made by Renaissance scholars. These scholars based their meteorology on the first three books of Aristotle's *Meteorology,* which was concerned not with forecasting but rather with attempting to give an account of change in the sublunary region.

The first three books of the *Meteorology* discuss changes in the sublunary region, a region that, according to Aristotle, is filled with irregular and episodic changes. This irregularity is in contrast to the celestial regions, where planets and stars, composed of an ethereal element, move eternally in perfect and circular motion. Sublunary change results from the eternal motions of the celestial bodies that drive the transformation and cyclical motions of the four elements. The motion of these elements is either toward or away from the earth's center and, due to the imperfection of matter, the motion is irregular. Aristotle described the proximate cause of meteorological phenomena as being two exhalations that move in continual cycles between the surface of the earth and the uppermost limit of the terrestrial region. The movements of the two exhalations provide a unity of explanation for Aristotle. They give an account for a wide variety of phenomena, including many that are now considered to be beyond the scope of the atmospheric sciences, such as the fiery paths of comets and the flickering light of the Milky Way. An analogous pair of exhalations circulates beneath the surface of the earth and explains geological and hydrological phenomena such as earthquakes, hot springs, the formation of metals, and the features of seas and rivers. The employment of the two pairs of exhalations as the material causes of sublunary change was long lived and endured even in the works of self-described opponents of Aristotle. Descartes, for example, appealed to these exhalations, despite dismissing the Aristotelian division of supralunary and sublunary, or in other terms, celestial and terrestrial physics.

For Aristotle and his followers, the motions and circulations of combinations of the exhalations are the unifying causes of most meteorological phenomena. One of the exhalations is vaporous; it comes from and is like water in terms of its powers. The other is hot and dry and similar to fire (1.3.340b27–30). The exhalations are similar to Aristotelian elements, in that they are not found in a pure form in nature but are always together. Precisely how the two exhalations combine is unclear, but presumably they form a combination made out of a juxtaposition of parts rather than a true homeomerous mixture. In any case, the predominate exhalation determined how it should be identified (2.4.359b32–34). Thus, an exhalation that Aristotle called vaporous contains both the vaporous and the hot and dry exhalations, even though the vaporous exhalation predominates.

The two exhalations are described as causing meteorological phenomena in several ways, according to Aristotle, whose descriptions of the exhalations are similar to those found in corpuscular explanations. For vapor, the primary mode of change is through the processes of composition (*sunkrisis*) and evaporation (*diakrisis*). The heat generated by the motion of the sun breaks water into smaller parts, thereby causing vapor to rise. After the vapor reaches the upper regions of air, the cold turns it into a liquid, causing it to fall. The intensity of the cold determines whether the vapor coalesces into hail, snow, or rain (1.9.346b21–35). Thus vapor (*atmis*) has different characteristics depending on the size into which it is condensed or rarefied (1.9.347a8–13). Elsewhere Aristotle had dismissed Democritus's attempts to explain the creation of new substances through the processes of *sunkrisis* and *diakrisis*, arguing that the results of these processes would merely result in a juxtaposition of particles, that is, a composition rather than a truly homeomerous substance.[11] While a critic of Aristotle might accuse him of lacking consistency, a more charitable approach would emphasize that the exhalations are not in fact true mixtures, because mixtures are homeomerous substances. As a result, they can be explained by the juxtaposition of particles or through *sunkrisis* and *diakrisis*.

The actions and the understanding of the subspecies of vapor are relatively simple compared to those of the hot and dry exhalation. Aristotle confessed that "there is no name that applies to it as a whole, and we are compelled to apply to the whole a name which belongs to a part only and call it a kind of smoke" (2.4.359b30–32). Just as there is no good name for the category as a whole, "there is no common name for all of its subspecies" (1.4.341b15). At different times, Aristotle called it smoke (*kapnos*), wind (*pneuma*), or fire (*pur*). Even though this exhalation was not to be identified with elemental fire or flame, he considered fire to be a suitable name for this stuff because of its inflammability (1.4.341b16–18). Aristotle envisioned that this substance fills the outermost part of the sublunary region, the zone closest to the sphere of the moon. This outermost region is volatile and bursts into flames when provoked by the smallest motions. Another subspecies of the dry exhalation is the matter that forms winds (2.4.360a15). For Aristotle, winds are not simply air in motion but are specific flows analogous to rivers. They have directions, sources, and seasons. Their matter is not elemental air, as one might expect, but rather the smoky exhalation. Fire-winds, hurricanes, thunderbolts, and lightning can also form out of this exhalation, depending on the fineness of the texture of the *pneuma* (3.1.370b5; 3.1.371a30).

Many of Aristotle's descriptions of the ways both exhalations create weather phenomena rely on the position and shape of matter, thereby evoking corpuscular motifs. The different characteristics of shooting stars depend on the position (*thesis*) and quantity (*plethos*) of the dry exhalation (1.4.341b24). When broken into small particles (*kata mikra*), the hot and dry exhalation can create sparks, which Aristotle referred to as a "goats" (*aiges*; 1.4.341b30). The constriction and concentration of cold expels heat, a process called antiperistasis, which in turn ignites the exhalation, creating shooting stars and similar phenomena. This force emanating from cold vapor is likened to pressure, as when the pressure applied by fingers shoots out a small object that was held between them (1.4.342a1). A nearly identical explanation applies to thunderbolts and hurricanes (2.9.369a25). Aristotle hypothesized an alternative explanation for shooting stars that primarily relies on the position and disposition of the hot and dry exhalation: instead of being the result of the motion of a single body, a shooting star is a chain reaction of sparks that follows along the path of a line or arc (2.4.342a1).

Aristotelian elements possess varying degrees of lightness and heaviness, which are the principles that direct their natural motions. Heavy objects, such as those whose composition is dominated by earth and water, naturally move downward, that is, toward the center of the earth. Air and fire are naturally light and thus move upward, or away from the center of the earth. A critic of the Aristotelian system might contend that the elements over time would separate as they are drawn to their natural place, leaving behind a world that is completely stratified and without any mixtures of the elements. The precise solution to the conundrum is perhaps unattainable, although attempts to solve it abound: substantial forms act as unifying principles to mixtures, fatty moisture acts as temporary glue to the ingredient of mixtures. Nevertheless, the elements partially separate, forming strata predominately, but not completely, composed of one element.

The elements' tendency to separate into distinct strata led followers of Aristotle to interpret *Meteorology* 3 as dividing the area between the surface of the earth and the moon into a fiery region, immediately below the celestial region, and an aerial region, between the fiery region and the earth's surface. The fiery region, which is composed not entirely of the element fire but also of water and other elements, is heated by the motions of the heavens. The aerial region has three distinct strata. The highest region, which was referred to sometimes as *aestus*, was thought to be hot and dry because of its proximity to the fiery region. Its heat rendered it inhospitable to clouds, which, according to standard Renaissance views, do not form above the height of mountains. Rather clouds were believed to form only in the lower parts of the aerial region, which were hot and wet, because of the circulation of vapor and the hot exhalation. Most authors of Aristotelian meteorological treatises believed that the lowest region of air was warmer than the middle region, a fact confirmed by the experiences of those who have travelled to higher elevations and by the usual absence of clouds close to the surface of the earth.[12] The heat of the lower aerial

From Oronce Finé, *Protomathesis* (Paris: Morre & Pierre, 1532), 103r. Courtesy of the Huntington Library.

region was attributed to the reflection of the sun's rays off the earth's surface (1.3.340a31). A woodcut from Oronce Finé's *Protomathesis* gives a schematic depiction of the strata, showing the formation of clouds only in the middle region.

THE DEFINITION OF METEOROLOGY

Renaissance authors of meteorological treatises typically defined meteorological phenomena in several ways. Frytsche's textbook provides four definitions for *meteora*. The most common definition during the early modern period was used by Descartes as well as by nearly all Aristotelians: "A *meteoron* is a body imperfectly mixed, generated out of vapor or exhalation in the region of air by the force and heat of the heavenly rays." [13] The designation *imperfect mixtures* (*imperfecta mixta*) must be understood in terms of the elements and their combinations. According to Aristotelian natural philosophy, four elements (earth,

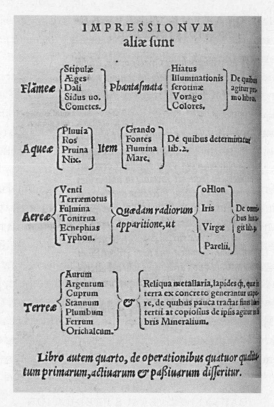

Above and opposite, from Hieronymus Wildenberg, *Totius naturalis
physicae in physicam Aristotelis epitome* (Basel: Oporinus, 1548), 190.
Courtesy of the Huntington Library. Translation by Craig Martin.

an integral part of university curricula, but also included the larger literate population, who purchased annual almanacs and ephemerides and read announce-
ments that predicted floods.[18]

A number of Renaissance writings on weather forecasting based their authority on ancient sources. A work on weather signs attributed by some to
Theophrastus of Eresus, Aristotle's student and heir at the Lyceum, cataloged
rules that were aimed to predict the weather. Some of the rules were based on
the position of celestial bodies, but most were based on observations of the
skies and of natural beings. For example, it was thought that "thunder in winter
at dawn is a rather good sign of rain," and "if a finch sings at dawn in an inhab-
ited house," it is a sign of rain.[19] Aratus's *Phaenomena*, a work dedicated to links
between weather and the position of stars, was also available and influential
during the Renaissance. Although similar traditions of predictive maxims arose

These Are Different Impressions

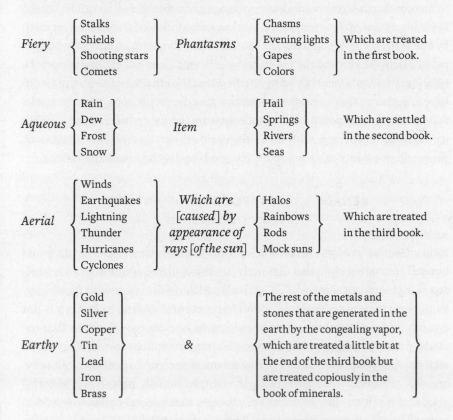

Fiery	{ Stalks / Shields / Shooting stars / Comets }	*Phantasms*	{ Chasms / Evening lights / Gapes / Colors }	Which are treated in the first book.
Aqueous	{ Rain / Dew / Frost / Snow }	*Item*	{ Hail / Springs / Rivers / Seas }	Which are settled in the second book.
Aerial	{ Winds / Earthquakes / Lightning / Thunder / Hurricanes / Cyclones }	*Which are [caused] by appearance of rays [of the sun]*	{ Halos / Rainbows / Rods / Mock suns }	Which are treated in the third book.
Earthy	{ Gold / Silver / Copper / Tin / Lead / Iron / Brass }	&	{ The rest of the metals and stones that are generated in the earth by the congealing vapor, which are treated a little bit at the end of the third book but are treated copiously in the book of minerals. }	

In the fourth book, the operations of the four primary active
and passive qualities are discussed.

autochthonously in all premodern societies, some Renaissance scholars, perhaps most notably Agostino Nifo, were influenced by their esteem of antiquity and considered the Theophrastean treatise to be a privileged guide to weather prediction.[20]

These two meteorological traditions are by and large divided. Forecasters or those concerned with the theory of forecasting were mostly empirical in their approach and ignored the meteorological causes behind the supposed correlations. Rather they were devoted to determining the accuracy of the signs. In contrast to these works, treatises on the causes of meteorology, which reflected the instruction of meteorology in universities and adopted the two exhalations, referred only infrequently to the practical concerns of astrology or the reality of the metaphysical teachings found in Aristotle's *Physics*, such as matter and form, potency and act, requires instruction. The primary material explanation of Aristotelian meteorology, the dual exhalations, fits with basic experiences of clouds, fog, and mists. Thus Aristotelian meteorology was, at least on some level, accessible, in addition to being a concern of nearly everyone. These qualities made meteorology an appropriate topic to introduce natural philosophy to those without formal university educations during the Renaissance. Moreover, the effects of the weather on numerous crafts made meteorology a practical concern of sailors, physicians, and architects.

Although meteorology might be accessible, weather itself is mysterious, at times awe-inspiring, and in many locales unpredictable. Its unpredictability combined with its potential for violence and its omnipresence has long sparked human curiosity. The earliest Greek philosophers and sophists discussed the causes of the weather; Aristophanes' satire of Socrates characterizes him as a *meteorologikos* lost in the clouds.[27] Even after the Presocratics replaced the divine will found in Greek myths with natural causes—Thales contended that movements in the water on which the earth rests cause earthquakes, not Poseidon—connections between the weather and religious understandings of the world remained common.[28] During the Renaissance many thinkers attempted to place meteorological catastrophes and rare events in theological or religious frameworks, discussing them in relation to biblical narratives, such as the universal flood, or as portents, signifying God's will. Even though daily weather is notoriously unpredictable, the broad strokes of meteorology—climate and the passing of the seasons—are regular. That regularity, on which growing seasons and harvests depend, suggests a cosmic order and for some thinkers, Stoics and Christians alike, divine providence. Providence, while a key concept in Christian thought from Augustine on, became more significant in natural philosophical settings during the Reformation, especially after Philipp Melanchthon used the concept as the centerpiece for his reforms of Lutheran universities.[29] At this point, natural philosophical understandings of meteorology became intertwined with the religious meaning and significance of weather itself.

METEOROLOGY AND RENAISSANCE ARISTOTELIANISM

This study supports and intends to expand on Charles B. Schmitt's argument that there were multiple Aristotelianisms.[30] Aristotelian meteorology differed among courtly elites, Italian university professors, members of Catholic religious orders, and Lutherans. Nevertheless, despite the variety of thought, certain elements resonate with larger trends in Aristotelian thought in particular and Renaissance intellectual life generally.

The use of Aristotle's writings as a foundation for natural philosophy was not new to the Renaissance. The Aristotelian corpus had been considered nearly comprehensive in its treatment of nature in several earlier periods and in other historical contexts. Aristotle's works put forth a broad framework for investigations into nature. Their grand scope, coupled with their general internal coherence, made Aristotle's writings the cornerstone of natural philosophy. The initial books of the *Physics* establish the first principles of investigations into nature: matter and form, potency and act, definitions of the artificial and the natural, and the four types of causes (material, efficient, formal, and final). The subsequent books of the *Physics* and the rest of his natural philosophical works apply these principles in their investigations of the cosmos as a whole, in the *De caelo*, and down to its smallest parts, in biological works such as *De partibus animalium*. In late antiquity, the Neoplatonic schools at Alexandria and Athens appropriated these works, along with Aristotelian logic, writing commentaries and giving lectures as propaedeutic studies to the metaphysics of Plato's *Parmenides*.[31] Later, Aristotelian natural philosophy became integrated with Arabic *falsafa*, beginning with the works of tenth-century translators and continuing in the works of Avempace, Avicenna, and Averroes, among others. After the reintroduction of Aristotle's writings on natural philosophy into the Latin West in the twelfth century, his opera became fundamental for investigations into nature until the end of the Renaissance.

The large body of writings that adopted, discussed, and grappled with Aristotelian natural philosophy in these fifteen hundred years is found in diverse genres of literature. These genres include simple translations, paraphrases, thematic treatises, encyclopedias, schoolbooks, scholastic *quaestiones*, and commentaries. In these writings, scholars often devoted themselves to explaining Aristotle's text. This was an onerous task due to its inherent philosophical difficulty, scholars' problems with translating and understanding Greek terminology, and Aristotle's notorious obscurity.[32] For many, properly understanding Aristotle's intent was a path toward a greater understanding of science. Thus, exegesis was a method for scientific discovery. Nevertheless, since the substance of an exposition of a text is dependent on the interpreter's cultural and intellectual position, commentaries often did more than restate Peripatetic doctrine. The very act of interpreting Aristotle's text was an interpretation of nature. Thus commentaries and *quaestiones* were both depositories of and paths toward knowledge about the natural world.[33]

Aristotelian natural philosophy was relatively conservative. Since many thinkers from Greek antiquity through the Renaissance shared similar goals, that is, to interpret Aristotle and thereby interpret nature, changes in the tradition came slowly and infrequently. Some commentators, most notably Alexander of Aphrodisias, Averroes, and Thomas Aquinas, remained influential for centuries. Many commentators were familiar not only with Aristotle's writings but also with the interpretations of their predecessors and thus frequently did not attempt to present an original interpretation of the text but rather reconciled earlier interpretations or selected what they saw as the best interpretation already available in the intellectual domain.

To divide the meteorological commentary tradition of the sixteenth and seventeenth centuries from that of the preceding centuries is admittedly artificial, but only to a point. There is much continuity within the entire commentary tradition. Both medieval and Renaissance scholars wrote their commentaries and *quaestiones* while employed in universities, which for the most part witnessed no radical changes in curriculum. Because both medieval and Renaissance treatments of meteorology largely arose out of similar institutional settings, they possessed similar goals, that is, they sought to explain Aristotle's text to students and thus were often versions or transcripts of public lectures and disputations. Furthermore, Renaissance writings did not totally replace medieval ones. Several medieval commentaries went through multiple editions in the sixteenth century, although the frequency of printing these works subsided by the end of the century. In the Renaissance, the most influential medieval commentaries on the *Meteorology* were those of Averroes, Albertus Magnus, Thomas Aquinas, and the *Quaestiones* of Themo Judaei, all of which were more than occasionally read, judging from the citations contained in treatments of the *Meteorology*.

There are, however, valid reasons to separate sixteenth-century meteorology from that of earlier centuries. The emergence of an Aristotelianism conditioned by humanist ideals marks the broadest difference between medieval and Renaissance thought. On the whole, sixteenth-century authors of meteorological tracts self-consciously distanced themselves from the medieval tradition. Humanist philological ideals conditioned readings of Aristotle during this time period. Some scholars read the original Greek and some concerned themselves with attempting to find historically accurate views of ancient authors without regard to whether the ancient opinions were true or not. These authors also reflected developments in humanism by utilizing newly discovered ancient sources. Renaissance humanists' emphasis on collecting, discovering, and analyzing ancient writings led to a growth in the number and diversity of ancient views available to Renaissance thinkers. Although Aristotle's writings were dominant in their influence, the availability of other ancient works led to the consideration of a multiplicity of views of meteorology and at times to eclectic or syncretic positions. These other writings included Plato's *Timaeus*; the works of Roman authors, such as Seneca, Pliny the Elder and the Younger, and Lucretius; the Greek commentators on Aristotle; and later Peripatetic works, namely Theophrastus's *De ventis* and the *Problemata*. Seneca's *Natural questions* looked at marvelous and wondrous examples of meteorology as part of a larger project of ethics. His considerations of the marvelous were intended to demonstrate divine providence, and his discussions of natural disasters put forth the thesis that because catastrophes can potentially occur anywhere, we are all at risk. Peace of mind can only arrive after a philosophical assessment of mortality. Thus Seneca's views, along with Pliny's description, influenced sixteenth-century discussions of earthquakes, volcanoes, and other violent manifestations of nature, all of which were considered meteorological at the time. Meteorological accounts by Epicurus and by Lucretius used multiple indeterminate causes and also found their way into discussions of some of the most intractable meteorological subjects, before becoming the inspiration for Pierre Gassendi's natural philosophy in the middle of the seventeenth century.

Particular events and discoveries of the Renaissance also changed aspects of meteorological thought. The discovery of the new world led to reevaluations of Aristotelian positions of climatic zones. The invention of siege mines suggested new explanations for the material causes of earthquakes. Natural events also directed research. The emergence of a new mountain on the sulfuric-smelling *campi flegrei* in Campania prompted Simone Porzio to consider chymical causes for subterranean disruptions.[34] Controversies arose between secular and religious authorities about the nature of frightful earthquakes that shook Ferrara during the 1570s, which resulted in the publication of numerous treatises and dialogues that investigated the causes of those tremors. The apocalyptic views of some Protestants combined with the astrological prognostication of a universal flood in 1524, which was unrealized, spurred discussions of the physical causes of flooding and the possibility of natural universal floods. An actual flooding of the Tiber in 1557 moved engineers such as Andrea Bacci and Antonio Trevisi to examine the causes of inundations in order to find ways to protect cities from them.[35] Thus Renaissance meteorology was informed by both its intellectual climate and the actual weather of the sixteenth century, giving it a more flexible character than both medieval meteorology and other fields of natural philosophy.

That this study, despite considering meteorological discourses written throughout Europe, has its center in Italy is not accidental. The nature of Renaissance Aristotelianism often leads discussions to Italy. The role played by Italians in this book corresponds to the available sources. During this period, Italians wrote far more commentaries on the *Meteorology* than did scholars in other European countries. Northern Europeans were more likely to write paraphrases, textbooks, and short commentaries than the Italians, whose works often left few issues untouched. As a result, Italians were more likely to discuss in explicit terms their motivations and influences as well as to comment on a given issue. Furthermore, Italy, despite its fair weather, was and continues to be the location of many natural disasters, making it an important place for discussions of these catastrophes during the sixteenth and seventeenth centuries. Natural philosophy and Renaissance Aristotelianism, however, were part of an international culture; thus an accurate portrayal must consider the meteorological tradition in a pan-European context. Attacks on Peripatetic philosophy were also pan-European, and so the later chapters, dedicated to the meteorologies of the emerging novel natural philosophies of the seventeenth century, increasingly integrate material from both below and above the Alps.

An examination of Renaissance meteorology gives a new perspective to Aristotelian natural philosophy. Many of the stereotypical perceptions of scholasticism simply do not apply. Authors of meteorological treatises recognized that understandings of this field should be flexible and limited. Meteorology was believed to be practical and part of public life. Moreover, Aristotelian meteorology was successful in its explanations. The concepts of imperfect mixtures and exhalations lived on in the work of some of Aristotle's sharpest critics. The advent of corpuscular philosophy in the first years of the seventeenth century corresponded to alterations of Aristotelian natural philosophy from within that used the *Meteorology* as the basis for an experientially based corpuscular philosophy.

CHAPTER ONE

The Epistemology of Meteorology

The typical career trajectory in the Renaissance universities of Italy made it natural that professors who wrote on meteorology would consider the field with respect to epistemology. Most professors began their careers teaching logic. Those who advanced often attained more prestigious and higher-paid positions teaching natural philosophy, possibly before becoming professors of medicine.[1] Thus throughout the sixteenth century, many of the most prominent authors of meteorological treatises had backgrounds in logic. At the beginning of the century, for example, Agostino Nifo's commentary on *De interpretatione* was published in 1507, and his works on the *Analytics* came out in the 1520s, roughly at the same time his commentary on the meteorology was published. Pietro Pomponazzi spent only a small part of his career lecturing on logic but nonetheless lectured on *De interpretatione* in the 1490s, approximately thirty years before turning his attention to the *Meteorology*. Similar patterns hold in the latter half of the century. Both Francesco Piccolomini and Giacomo Zabarella, two of the most famous professors of natural philosophy at Padua during the sixteenth century, taught logic before they wrote about meteorology.[2]

Their familiarity with syllogisms, epistemology, and what more recent commentators have called *method* led natural philosophers to consider the status of knowledge of meteorological phenomena. Realizing that Aristotle's logic offers a number of paths toward understanding, authors of meteorological treatises recognized that the *Analytics* presented idealizations of knowledge that were not always applicable because of limitations to human understanding and the irregularity of some facets of nature. These discussions of the status sublunary projectiles, is a sign (*semeion*) that they are sublunary.[27] In *Meteorology* 2.2, Aristotle maintained that an artificially contrived test, in which he claimed that eggs float on water that contains salt, is a *tekmerion* that demonstrates that fluids become denser when mixed with other substances.[28]

Perhaps Aristotle's most extended use of signs is found in the explanation of earthquakes in *Meteorology* 2.8. In this chapter, Aristotle posits that vaporous winds moving underneath the earth's surface are the causes of tremors and quakes. Evidence for this theory comes from deduction, analogy, and signs. Aristotle deduced that wind must be the cause because wind naturally is the most powerful of all substances.[29] Analogies between the body and the earth suggest that pent-up air is the source of tremors and discomfort.[30] Signs based on sensation (*hemetere aisthesis*) provide further evidence that Aristotle's analysis is correct and winds are at the root of earthquakes. These signs (*semeia*) include secondhand observations of winds breaking forth from the grounds at earthquakes in Heracleia and the Aeolian Islands and generalized experiences such as the opinion that earthquakes occur when the sun becomes misty and dimmer despite the lack of clouds and that they often occur when the dawn is frosty and calm, all of which suggests that winds have left the aerial region and flown underground.[31] Moreover, the sound of subterranean winds in the Aeolian Islands that heralds southerly winds is an additional sign (*tekmerion*) that these subterranean winds exist.

The idea that natural signs were evidence of probable causes did not emerge newly in what Ian Hacking considered the "low sciences" of the Renaissance, such as alchemy and mining, and are not dependent on the Paracelsian concept of signatures.[32] Rather, arguments that used evidentiary signs played a continuous role in Aristotelian natural philosophy and are particularly prominent in meteorology because of the characteristics of the subjects of this field.

MATTER THEORY, UNCERTAINTY, AND METEOROLOGY

A number of Renaissance treatments of Aristotle's *Meteorology* not only accepted his claims that meteorological knowledge is limited but used matter theory to explain why. These explanations were based on two principles: (1) meteorological phenomena were imperfect mixtures, and (2) the matter of the terrestrial or sublunary realm was imperfect and, as a result, unstable and relatively unknowable.

The idea that the first three books of the *Meteorology* treated imperfect mixtures goes back at least to Albertus Magnus, who described the subject of these books as matter that is in the state of becoming a simple substance.[33] John Buridan was one of the first to use the term *imperfect mixtures* to categorize meteorological effects in contrast to the perfect mixtures, such as flesh, blood, milk, and metals, for which *Meteorology* 4 gives an account.[34] Renaissance Aristotelians followed Buridan's phrasing almost uniformly. The matter of meteorological phenomena was considered imperfect because it was seen as a composition of the four elements that had not fully mixed or transformed into something distinguishable from the elements with its own distinct substantial form. The acceptance of meteorological phenomena as "imperfect" contributed to the idea that meteorological knowledge is probable. If meteorology considers objects that are without their own substantial forms, knowledge of formal causes would be limited. Indeed meteorology most often referred back to the forms of the elements that composed the two exhalations, which were the material and efficient causes of meteorology. Francesco Piccolomini, a professor at Padua during the last decades of the sixteenth century, explicitly expressed this view: "The form that is a substance is not properly suited to meteors because they are imperfect mixtures, which are of such a type that they will not create new forms; therefore their substantial form is not distinct from those of the elements."[35] As a result, scholars were limited in the kinds of formal and final causation that could be used to understand meteorology.

Although Renaissance commentators did not uniformly deny there were formal or final causes for meteorology, the rejection of the existence of those causes was frequent. For example, Francesco Vimercati and Jacob Schegk specifically argued that the two causes of meteorological phenomena are efficient and material causes, and the Coimbrans' commentary contended that "meteors" do not have their own formal causes and made little mention of final causes.[36] Effectively, the idea that the end of a substance was the realization of its form was ruled out, leaving only external final causes available for those who believed that final causes do indeed exist for meteorology. The character of meteorological phenomena thus eliminated the possibility of a deep knowledge of their formal and final causes, the two most privileged types of causes for Aristotelians.

The Renaissance conception of the imperfection of meteorological phenomena was an amalgam of Aristotelian ideas of potency and act and Platonic views of matter, which had held currency throughout the Middle Ages. In the *Timaeus*, Plato contrasted the matter of the sublunary world with the forms. Forms are perfect, unchanging, and the source of knowledge. In contrast matter is the source of imperfection in the world because of its instability. Material necessity is the "errant" cause.[37] As a result, understandings of the material world are

The inherent difficulty of meteorology made its conclusions tentative until better principles could be uncovered.

Though he was a critic of a number of aspects of Avempace's natural philosophy, Averroes in his meteorological works agreed that meteorological knowledge has fixed limits. In his discussion of the Milky Way, Averroes argued that because there are doubts about the genus of the Milky Way, our knowledge of its causes should be accurately labeled "possible" and our understanding diminished (cognitio diminuta).[48] The acceptance of limitations would be no hindrance to speculating about these causes as long as we recognize that they are only possible; thus, Averroes offered two separate possibilities for the causes of the Milky Way's flickering: either the weakness of our eyes causes it because of the distance of small stars, or a place in the skies receives and multiplies the light of the stars.

The views of Averroes resonated during the Renaissance. Pietro Pomponazzi expressed skepticism about the possibility of complete knowledge of the natural world, using our inability to understand meteorological phenomena as evidence for the limitation of human knowledge. In his De incantationibus, a work dedicated to giving potential explanations for prodigies—strange and seemingly miraculous phenomena, including fountains and statues that drip blood or bizarre meteorological events chronicled in histories (the time it rained wool, for example)—he conceded that an epistemological standard below certainty was appropriate. In the introductory pages of De incantationibus, he contended that some Peripatetics used demons to explain these difficult-to-understand phenomena, not only because demons are posited by "ecclesiastical decrees" but also because their presumed existence allows us to "save many phenomena."[49] The employment of demons in natural philosophy is thus instrumentally justified, parallel to epicycles and eccentrics that save the phenomena in astronomy. Although Pomponazzi rejected demons as a cause in natural philosophy, he did not deny that naturalistic explanations are meant to save the appearances. Citing Averroes' commentary on the De caelo and Aristotle's Topics, Pomponazzi argued "that in difficult and hidden matters, the answers more removed from inconveniences, and more consonant to sensations and reason, are to be better received than contrary arguments."[50] As a result, Pomponazzi's controversial claim that the miraculous events recounted in scripture "can on the surface be reduced to natural causes" need not be taken as an endorsement that the events actually were the result of natural causes alone but rather can be taken as support of the slightly more modest claim that natural causes can give an explanation that potentially conforms to our experiences and reasoning.[51]

Pomponazzi further considered epistemology and the natural world in his *In libros meteororum* (circa 1522). In this book of *quaestiones*, he contended that Aristotle at times adopted the epistemological standard of verisimilitude and employed rhetorical arguments in natural philosophy. As a result, Pomponazzi conceded that saving the appearances was an appropriate ideal for natural philosophy.

In a *dubium* dedicated to *Meteorology* 2.1, he addressed the degree of knowledge and certainty that can be applied to meteorological subjects as he tried to make sense of a passage of the *Meteorology* that confounded and contradicted his own experience of the natural world. In this chapter, Aristotle attempted to explain the sources that create various kinds of bodies of water. He divided the kinds of bodies of freshwater into two types: standing and running. Running water, such as rivers and streams, comes from sources that are higher than the stream or river. Standing water, however, is of two types. Typically, standing water naturally comes from the accumulation of rainwater and is static; the relevant examples are lakes and swamps. According to Aristotle, standing water that comes from underground sources only does so when it is artificially created, as in the case of wells. This last statement, which Pomponazzi found problematic, does not appear to have been a slip of the Stagirite, as it appears twice. Aristotle wrote, "Some [standing water] springs from sources, but are always made to do so artificially (*cheirokmeta*), as for instance the water in wells." And two lines later, he wrote, "Hence water in streams and rivers runs of its own accord (*automata*), but well-water needs an artificial construction (*technes ergasomenes*)."[52]

Pomponazzi found untenable the contention that all standing springs or wells must be artificial: "This is *contra experimentiam*, as many from the schools have told me they have seen, and I myself have seen, many springs and natural wells."[53] After rejecting the interpretation attributed to Thomas Aquinas (which was shared by Gaetano of Thiene and "all of the Latins [he] had seen") because this view, although it was in agreement with Aristotle, was not in accordance with reason and experience, and after dismissing Alexander of Aphrodisias's view, which according to Pomponazzi was also false, he tried to determine why Aristotle believed as he did. He speculated that perhaps "in Aristotle's country all stationary bodies of water are man-made," although in his own regions "this is not the case."[54] But his conclusion was that Aristotle's view is "probable and does not demonstrate."[55] More strongly he contended that "in my judgment Aristotle's theory is without value."[56] The only way Pomponazzi found to make sense of this passage was to lower the epistemological standard that

A similar skepticism toward the possibility of certain knowledge of meteorology is found in Tiberio Russiliano's *Apologeticus adversus cucullatos* (1519), a work that applied physical causes to both the foundations of Christian dogma and to what was traditionally understood as miraculous. Russiliano, a former student of Nifo, in a *quaestio* in which he put forth the contention that according to philosophical arguments there must have been an infinite number of universal floods, considered his contention "demonstrative and unassailable," but only if his suppositions are accepted. He admitted that not all of his suppositions were necessarily true even if they are clear in themselves (*ex se patent*); rather they were conjectural, based on common agreement and sensible signs: "The suppositions are clear in themselves; first they derive out of common agreement and experience, since signs and traces of inundations appear in mountainous regions, such as seashells and oysters, so that we should reasonably arrive to the conjecture that when there was a universal flood it covered and surpassed all of the mountains."[65] The key is that signs and traces, based on experience, lead to conjectures that then become the basis for premises in deductive arguments. The deductive arguments, however, are only as sound as the conjectures upon which they are based.

Scholars outside of Italy held similar views as well. Jacob Schegk is the most prominent example of a Lutheran scholar who argued that the epistemological standard of meteorology was similar in its *akribeia* to mathematical fields. Having cited the first book of the *Ethics*, Schegk wrote that for many parts of nature we cannot trust our solutions but rather we should regard our beliefs with less certainty.[66] This lower standard was especially applicable to investigations into phenomena that arise contingently out of matter, as is the case for meteorology. In those cases Schegk contended that we should expect our explanation to not explain what is necessarily the case—because we cannot know it—but rather we should hope to give an account of what is possibly the case.[67]

Since the time of Pierre Duhem, the epistemological standard of "saving the appearances" has been used to distinguish the astronomical science from cosmology and terrestrial physics.[68] As Peter Barker and Bernard Goldstein have argued, the epistemological goal of "saving the appearances" was meant to give *demonstratio quia* while admitting the impossibility of *demonstratio propter quid*.[69] While for astronomy the limitations on certainty result from the mathematically indistinguishable nature of certain principles and the inability to observe accurately celestial bodies, for meteorology the limitation also derives from the accidental nature of meteorological phenomena. Lodovico Boccadiferro, a professor of natural philosophy at Bologna from 1527–45, who had studied with Pomponazzi, perhaps gave the clearest summation: "This law must be observed: that when the causes of some effects are unknown to us, we must accept suppositions, or principles, from which nothing impossible, nothing contrary to the senses, and nothing repugnant to the appearances follows." Boccadiferro admitted that this "contingent possible proposition is that which is not true, but could be true."[70] Thus the epistemological standard for meteorology fell far below that of certain truth, and the goal of "saving the appearances" should not be taken to be characteristic only of astronomy but for some parts of natural philosophy as well.

OBSERVATIONS, THEORY, AND REVISIONS

The contention that meteorological theory was conjectural and at times only capable of saving the appearances was widely known among those concerned with this field in the sixteenth century. Moreover, this contention justified the application of new observations, which could be used as signs to correct Aristotle's own theory. For example, a number of Aristotelians as early as the 1520s used the observations of sailors to amend Aristotle's position that there was an uninhabitable torrid zone in the area around the equator.[71] Moreover, Vimercati contended that, contrary to Aristotle, Portuguese sailors and Columbus had observed flows in the Atlantic Ocean, which had caused their return trips to be of different lengths than their departing voyages.[72] Zabarella used his experience of being upon Monte Venda, outside of Padua, on a day when it rained in the lowlands but did not on the mountaintop to conclude that he had observed the "middle region" of air, that is, the region above the clouds. From his observations, he concluded, against Aristotle, that this region is composed of normal air, not exhalations.[73]

The tradition of using experiences to correct Aristotle continued in the work of Niccolò Cabeo, a Jesuit originally from Ferrara, who emphasized experience and chymical experimentation in his 1644 commentary on Aristotle's *Meteorology*.[74] He noted approvingly that Aristotle used the first half of the *regressus* method in the analysis of earthquakes: "He [Aristotle] began well, as I have said, in trying to show a posteriori, or rather by the *methodus resolutoria*, what is the cause assigned to earthquakes."[75] For Cabeo, however, Aristotle's attempts were insufficient, and his theory was unable to explain the real cause of earthquakes. Experience and observations taken from recent earthquakes support a chymical explanation. Instead of maintaining that winds or the eruption of subterranean exhalations provoked tremors, Cabeo believed that earthquakes result

CHAPTER TWO

Teleology in Renaissance Meteorology

The irregularity that contributed to the lack of certainty in meteorology also contributed to the rise of discussions of the purpose of meteorological phenomena. There is little if any discussion of teleology in Aristotle's *Meteorology*. The apparent absence of the consideration of final causes, while shifting much discussion to material and efficient causation, led some natural philosophers to question whether the Aristotelian doctrine could not be improved by inventing or discovering final causes, which Aristotle himself did not describe. They asked whether even if meteorological phenomena were accidental, composed of imperfect mixtures, do they not have a larger purposive role in the universe or in human affairs?

The divide between the prominence of final causes in Aristotelian natural philosophy and the rejection or severe limitation of final causation as an acceptable explanation of the natural world by figures, such as Bacon, Descartes, and Spinoza, during the seventeenth century has been considered a distinguishing mark between premodern and modern science.[1] Nevertheless, proponents of the mechanical and corpuscular philosophies of the seventeenth century were not all opponents of teleology. Pierre Gassendi and Robert Boyle endorsed teleology, Leibniz embraced entelechies, and they creep into Descartes' psychology, despite his adamant attempts to eliminate them.[2] Nonetheless, critiques of ends in natural philosophy resonated throughout seventeenth-century natural philosophy and beyond. Enlightenment figures such as Jean Le Rond d'Alembert ridiculed the use of teleology to explain the natural world.[3] Dennis Des Chene, however, has effectively demonstrated that Descartes' and others' characterizations of early modern Aristotelians as positing intention into nonrational or even nonsentient agents did not reflect their true position.[4] Des Chene's view holds for sixteenth- and seventeenth-century Aristotelians, which he treats with much skill and subtlety through exploring philosophical arguments mostly found in discussions of *Physics* and *Metaphysics*.

An examination of teleology in late Aristotelian meteorology adds to these studies by showing the extents to which final causes could be applied to the natural world. Consideration of the teleology of meteorology illustrates the relationship of different camps of Aristotelians to religious concerns, the purposes of universities, and the methods of reading Aristotle. Debates over teleology were especially evident in considerations of meteorology, a field that demanded causal explanations of irregular and at times intractable phenomena. This field did not consider souls, ordered nature, or, according to some interpretations, even substantial forms. As a result the field is particularly pertinent to questions over the nature and extent of final causation in Aristotle, issues that are much debated in contemporary scholarship. Similarly, Renaissance Aristotelians debated the role of final causes in natural philosophy and perhaps had even more divergent views on the topic than those found in modern interpretations of Aristotle. Nevertheless, early modern concerns found in treatments of meteorology partially correspond to recent discussions, some of which assert that, according to Aristotle, seasonal rains are teleological, perhaps in an anthropocentric sense, while others contend that Aristotle's teleology is limited to natural kinds.[5]

The doubts over final causes in meteorology have been and are heightened by the lack of discussion in Aristotle's *Meteorology*. While the necessity of rain for agriculture might obviously point to an anthropocentric purpose, the ends of meteorologically provoked disasters are less clear or even dubious. What is the purpose of a destructive storm or flood? Answers to these questions were inevitably linked to theological and ethical concepts during the Renaissance, and the character of these answers in the sixteenth century largely differed according to the confessional divide between Protestants and Catholics.

The difficulty and uncertainty engendered by meteorology made it possible for authors from Italy, influenced in part by medieval theology, to contrast the limitations of human knowledge with God's absolute power. For them, the difficulty was evidence of the inability of humans to understand the purpose of the accidental, contingent properties of matter. While this inability could cut short discussions of the teleology of meteorology, some natural philosophers took this as evidence of the nondeterministic character of the natural world, which ultimately depends on God's will. Divine power extends not just to substances but to accidents as well through secondary causes. Even if it is not possible to grasp God's precise intentions, these accidents are purposeful because they are part of the divine ordering of the universe.

A number of sixteenth-century Lutheran philosophers expounded on the final causes of the weather in lengthy and sophisticated discourses with greater certainty than did most professors working in Catholic lands. Lutheran philosophers often saw meteorological phenomena as proof of divine providence or God's wrath. In a reworking of Aristotelian concepts of teleology, final causes were evident not only in the necessary links between benign weather and human welfare but also in the seemingly portentous nature of rare and violent events. Lutheran authors saw meteorological events as prophetic signs, which found their purpose in foreshadowing the future.

Lutherans' use of final causes emerged from their self-conscious attempt to fuse natural philosophy with theology. Philipp Melanchthon, who was responsible for creating the curricula of Lutheran universities in the 1520s, emphasized that the teaching of natural philosophy should show God's providence and the ordering of the universe.[6] These goals dominate later works, such as those by Nicolaus Taurellus, who, in his book-length attack on the Italian professor of botany and medicine Andrea Cesalpino, disputed Cesalpino's contention that *meteora* do not have final causes.[7] In contrast, Taurellus thought the failure to see purpose in nature was emblematic of the kind of natural philosophy practiced in Italy often by laymen (as most Italian professors were), who considered natural philosophy propaedeutic to the study of medicine and largely distinct from theology.[8] Even if many of Taurellus's views on philosophy were atypical, views similar to his on teleology were widespread but not universal among Lutherans. For example, Jacob Schegk and Georg Liebler, both of whom taught natural philosophy at Tübingen, did not emphasize the examination of teleology in their treatments of meteorology. But at Wittenberg and Leipzig, Johannes Garcaeus and Wolfgang Meurer not only used final causes but also categorized the purposes of meteorology as either physical or theological. Physical final causes were typically anthropocentric or concerned with the functioning of the universe. Theological final causes were found in the supposed divine meaning of weather. Garcaeus and Meurer considered the purpose of rare or violent weather events to be found in their prophetic nature; they were signs of God's will, which could be at times providential and at times wrathful.

Although the teleology of natural philosophy was often related to religious issues, many Italian authors questioned whether meteorological phenomena were purposeful for other motives. Taurellus was correct in proclaiming that lay scholars working in Catholic lands, such as Italy, were more likely to dispute the applicability of final causes to meteorology. Their doubts, however, resulted not just from their lay status but also from their desire to attempt to interpret Aristotle's texts literally. Discussions of final causes for meteorology added little to a literal or historically minded reading of the text of the *Meteorology*. Nevertheless, factors such as the attractiveness of systematizing, considerations of meteorological phenomena as necessary parts of the universe, theological motivations, and open disagreement with Aristotle's texts prompted Italian natural philosophers to discuss the possibility of final causes for some meteorological phenomena.

TELEOLOGY IN ARISTOTLE'S *METEOROLOGY*

Aristotle wrote his natural philosophical works to attack, among others, certain Presocratics whom he considered to be materialist and determinist. Empedocles and Democritus failed, in his eyes, because they thought that the principles of necessity and chance were sufficient explanatory tools for all products of nature. For Aristotle, much of the content of the universe was clearly the result of design. Instead of utilizing the idea of a designer or craftsman, as Plato had done in the *Timaeus*, Aristotle linked the idea of design with nature, or *phusis*, which served as a directional agent. In his view, nature in general is the regulating principle of the universe. Each natural body also possesses its own *phusis*, which is its internal source of motion and growth. The completion (*teleiōsis*) of this motion, such as the growth of animals into adults or the natural motion of the elemental bodies into their natural places can be understood as a fulfillment of an intrinsic purpose. Cosmic order, nature's regularity, and fixed species of living things were taken as signs that the world was neither the product of chance nor merely the result of a random arrangement of atoms. The degree of order and regularity, however, was not uniform throughout the universe.

For Aristotle, the motions of the celestial bodies were the most orderly, moving perpetually in circles. Even though the motion of these bodies causes change in the sublunary realm, change is less ordered below the moon. In *De generatione et corruptione* 2.10, Aristotle argued that the motion of the sun along the ecliptic caused the elements to transform themselves cyclically. This motion is ultimately the producer of all generation and decay on earth. According to Aristotle, because the sun moves on the ecliptic it is thereby either approaching or retreating from the earth and "its movement will be irregular."[9] The impermanence and relative disorderliness of the sublunary realm is the result of the irregular motion of the sun. The proximity of the sun causes generation, while its remoteness causes destruction. Additionally, generation and corruption are irregular due to the matter of the terrestrial world: "For their matter is irregular. . . . Hence the processes by which they come-to-be must be irregular too."[10] Lack of order, however, does not entail a complete lack of perfection. Rather, according to Aristotle, God "fulfilled the perfection of the universe by making coming-to-be uninterrupted."[11] Therefore cycles of generation and destruction lead to the perpetual being and creation that participates in the perfection of the universe.

Although nature strives for the best, and purpose is found throughout nature, not everything is purposeful in itself, even if it is part of the general cyclical nature of the sublunary region. Rather some things are the result of material necessity, namely, what arises from the dispositions of matter alone. Aristotle gave perhaps the clearest example in *De generatione animalium* in his discussion of eye color. Eyes clearly have final causes since their proper functioning allows for sensation. The color of the eye is not always purposeful, however, because it typically is not related to the functioning of the eye or to the essence of the eye's possessor. In order to understand eye color, "We must refer the causes to the material and the motive principle on the view that these things come into being by necessity."[12] The accidental properties of the eye come out of a simple material necessity, an unqualified (*haplōs*) necessity, and the color itself has no purpose; it is not the result of the realization of the nature of either the eye or the animal. Consequently, knowledge of the eye's color comes only from an understanding of the properties of matter and cannot be placed within the larger context of the particular organism, its species, or the order of the universe.

Potentially, large portions of meteorological phenomena can be seen as accidental, that is, caused by material necessity alone. The first three books of Aristotle's *Meteorology* are almost if not completely dedicated to explanations via material and efficient causation.[13] Final and formal causes typically are not part of his explanations for these subjects because the matter of meteorological phenomena is perpetually imperfect, being partial transformations of the elements, as numerous medieval and Renaissance commentators noted. Moreover because these partially transformed elements are inanimate, as the sixth-century commentator Olympiodorus argued, they do not participate in the formation of organs and organisms, which have clear purposes and ends.[14] During the Middle Ages and the Renaissance, most Aristotelian commentators described the subject of the *Meteorology* as "imperfect mixtures." Meteorological phenomena were considered imperfect because they were composites of the elements that had not been transformed into a new substance. The forms of the elements remain, thus rendering meteorological phenomena without their own essential natures and substantial forms independent of the four elements. The imperfection of meteorological phenomena made discussions of their *teleiōsis* superfluous because they lack the intrinsic end of the realization of form. Although he did not rule out the possibility that meteorological phenomena have an extrinsic purpose, Aristotle failed to discuss what these extrinsic purposes might be. While many scholastics might have held that final causes only exist in relation to an intelligent agent, such as God, Aristotelians continued to find intrinsic final causes in the realization of the substantial form of inanimate bodies.[15]

Aristotle's followers in the Renaissance faced a dilemma. The failure of the *Meteorology* to discuss teleology, the irregular and even noxious character of many meteorological phenomena, and the understanding of it as being caused by imperfect mixtures suggested that the field of meteorology could dispense with final causes. At the same time, however, Aristotle could cite much evidence that there is some order and purpose to the weather. Seasonal rains ensure the availability of crops, and climatic and seasonal weather patterns both exist and are necessary for human survival.[16] The lack of consideration of final causes in the *Meteorology* does not necessarily mean that Aristotle outright rejected their existence. Moreover, Christian emphases on God's will, providence, and creation further complicated Renaissance discussions of the purpose of meteorology, since the idea that weather phenomena were without order suggested limitations both to God's power and to God's involvement with the mundane.

RENAISSANCE ITALIAN VIEWS ON ENDS
IN METEOROLOGY

While professors in universities in France, Spain, and other Catholic countries taught meteorology during the sixteenth century, the strength of Italian universities, with their emphasis on medicine and natural philosophy rather than on theology and with their teaching methods that depended on explication of Aristotle's text rather than on paraphrases, contributed to the production of some of the most detailed and sophisticated commentaries on the *Meteorology* in all of Europe during the Renaissance.[17] Moreover, the lay status of Italian professors coupled with their willingness at times to separate the study of nature from

Pomponazzi believed to be confirmed by experience: closed rooms are less healthy than airy spaces. Moreover, winds bring rain clouds, and farmers note that winds scatter seeds and cause grain to swell in size.

This analysis, however, does not end the discussion. What should be made of all the harm that violent winds create? One solution is to consider them to be failures to reach an end; they are nature's mistakes, such as when a "writer errs, a physician kills." In Aristotle's eyes, natural goals are for the most part, but not always, attained; nature's failure to reach perfection raises doubts. How is it that God created the winds for the greater good, but often they fail to assist that greater good and cause evil? God must know that this happens. Averroes' solution, that these bad events occur by chance and that only good things are known to God, is not satisfactory to Pomponazzi because it goes against Christian faith as well as against Aristotle's words. A partial solution is found in the intrinsic perfection of the meteorological phenomena; citing Pseudo-Dionysius Areopagita who contended that good dogs should be furious ones, Pomponazzi argued that a good wind is one that is strong, lightning that does not strike anything is not worthy of being lightning, and earthquakes are only good with respect to their genus if they destroy cities and provinces.[29]

The destruction of cities and their inhabitants, however, might seem to be a clear evil, not necessarily part of the functioning of a perfect universe. Pomponazzi's solution is surprisingly theological: "For God is the cause of all things, except evil desires, of which we are the cause. But there are many things that seem bad to us, which are optimal, because we are ignorant of their purpose."[30] Thus he argued that these destructive winds and earthquakes might seem bad, but we are not able to understand the purpose for which God created them. Echoing Aristotle's position outlined in *De generatione et corruptione* 2.10 that destruction is required for there to be perpetual generation and being, Pomponazzi gave examples that demonstrate the cyclical nature of destruction and death that in turn leads to generation. Many regions were required to become weak in order for Rome to have become an empire; for someone to become rich, many around him must become poor. He even suggested that it might be good "should the Turk come, because afterwards, we should be better Christians."[31] Conquest and humiliation could deepen faith.

Even though the belief in the unknowable nature of God's causes has its origins in theology, solace is to come not from Christianity but from philosophy, in particular Stoic philosophy. A realization that floods, earthquakes, and whirlwinds can strike anywhere could lead to a sense of lack of security. Security, however, according to Seneca, comes from the paradoxical understanding that

we are more secure when we admit that complete security does not exist anywhere. According to Pomponazzi, philosophers are those who are most secure
because they "know that everything happens from the order of nature, and
therefore they do not marvel at these effects as the unworthy vulgar do, since
they recognize the causes of its effects and that it is orderly and best according to nature. Therefore they know the positioning and *ordinatio* of God."[32]
Thus Pomponazzi revealed the ethical goals of natural philosophy: first, eliminating wonder by explaining the causes of natural effects; and second, bringing
further security through an understanding that God has ordered the world in
the best way possible.

The first of these two goals was at the heart of his *De incantationibus*, a work
in which he used hypothetical naturalistic explanations to suggest that wondrous and seemingly miraculous events could be explained by physical causes
rather than by demonic or angelic influences.[33] While some scholars have seen
this earlier work as evidence of Pomponazzi's alleged unorthodoxy, in the lectures on the *Meteorology* he maintained naturalistic explanations while also asserting the existence of an omnipotent God who is the ultimate cause of a well-
ordered universe. And although he appealed to Seneca's ethical justification for
the practice of natural philosophy, Pomponazzi's acceptance of a well-ordered
universe entails neither Stoic material determinism nor the belief that it is possible to know the causes for everything the universe holds. These positions lo-
cate Pomponazzi in a kind of middle ground, whereby he confirmed that the
world was divinely ordered and that contemplation of this order instills wisdom
yet still held that the vision of a complete philosophical understanding of the
world is the domain of fools.

The vain desire to know everything puts some philosophers in an unenvi-
able position: "My Lord, it is no wonder if philosophers are mocked by common
people, since they want to examine everything and what God can do . . . ; they
want to unite God's secrets and nature, therefore they spurn riches and plea-
sures." These philosophers, however, are mocked not just by common people
but by Pomponazzi as well. His assault better identifies the strains of thought to
which he objects, mainly Peripatetic proponents of astral determinism: "Peripatetics, however, and other stupid philosophers, who want to know everything, say that this [cyclical destruction and regeneration] happens out of the
necessity of the heavens." Accordingly, theologians offer a more attractive solution that admits the limits of human knowledge: "Therefore those *religiosi* do it
well and better, who respond securely, that because the will of God wishes it
such, thus it becomes such; therefore, no other cause of these things must be

false, that everything that happens, happens by necessity. Rather, neither God nor anyone, through God, understood this."[45] Pomponazzi, however, dismissed Averroes' conclusion, aligning himself with Thomas and asserting God's omniscience: "But you should note that D. Thomas and others say that this conjunction does not have a cause, although God establishes it, since he sees everything; and what S. Thomas says is most true, since nothing is simply by chance, since everything is known by God, and God has known and determined causes, even though according to nature this is by chance."[46] Pomponazzi distinguished between God's knowledge, which is perfect, and nature, which is imperfect and at times accidental. Attempts to understand these seemingly chance events by looking at nature are fruitless.[47]

Pomponazzi thus concluded by agreeing with Thomas, that God determines everything, while still maintaining that there are chance events, at least according to nature. As a result, events that occur by coincidence support the theologians who say that we cannot know all of God's purposes. Philosophy has self-recognized limits that can only be transcended by faith: "These are said by Theology and said well, but are not according to Aristotle's mind. Much more should be believed (credenda) here than investigated, but stupid philosophers wish to investigate everything. Aristotle wishes that the conjunction does not have a cause, and therefore that which these men [i.e., the stupid philosophers] say is neither Peripatetic nor Academic."[48] Efforts to increase the understanding of final causes in nature are unproductive because according to nature, coincidences do not have final causes. Only religious faith allows for the understanding that these purposes exist.

Not everyone was persuaded, however, that all meteorological effects possess some deeper purpose. Agostino Galesi, a professor at Bologna, writing in 1571 during a crisis caused by a series of earthquakes in Ferrara, pointed to the ongoing controversy over the teleology of such destructive events. His work, dedicated to the cardinal of Bologna, Gabriele Paleotti, dismissed the assertions that earthquakes are part of the natural order that brings about the perfection of the universe: "What then can be considered the perfection of such things, when they either portend evils or bring about these great calamities to men? Then truly, what is that perfection? Since in no way do they confer any order, or beauty, or utility to the universe . . . but rather they disrupt everything, strip away beauty, and demolish."[49] Galesi's tone evokes the urgency and despair felt in the aftermath of this disaster yet also reflects actual disputations that had taken place in Bologna, where he had earned his doctorate in arts and medicine and by 1571 had become a professor of philosophy.[50]

Lodovico Boccadiferro, also a professor of philosophy at Bologna, but a generation before Galesi, followed his teacher Pomponazzi, but only up to a point. He believed that the imperfect nature of meteorological phenomena means that we are ignorant of their final causes, even if such causes exist.[51] His view, however, still allows for the possible existence of final causes. In his commentary on the second book of the *Meteorology*, which was published posthumously in 1570, a year before Galesi's earthquake treatise and the actual year the tremors began to strike the Po Valley, the aptly named Boccadiferro used Aristotle's failure to identify final causes for meteorology as a starting point for a prolix discussion. For some meteorological phenomena his reasoning was similar to Pomponazzi's: its ultimate purpose was for the generation of being that accompanied the perfection of the universe.[52] In the case of earthquakes, however, he was less sure: "About the final cause I say that either the earthquake does not have a final cause, this effect arises out of the necessity of matter, the exhalation is not intended, and it does not intend this motion, but is according to the ascent [of the exhalation], or if it has an end, the end of the generation of the earthquake is because this motion of the smoky exhalation provides for that the realization of generated things, that is, the perfection of the universe."[53] While Boccadiferro gives a possible final cause, he also admitted that earthquakes might just be accidental effects caused purely from the "necessity of matter" lacking a purpose or utility to humankind or the functioning of the world.

LUTHERANS AND THE PROVIDENCE OF WEATHER

One line of Aristotelian thought, then, emphasized the limitations of meteorological knowledge. Lutherans, however, largely adopted a contrary approach and saw meteorological phenomena as legible signs of God's providence or anger. In a reworking of Aristotelian concepts of teleology, final causes were found not only in the necessary links between benign weather and human welfare but also in the portentous nature of rare and violent events. Meteorological signs were seen as prophetic, and their purpose was found in their ability to foreshadow the future. This view is reminiscent of Seneca's description of the Etruscans, a race famous as proponents of divination in antiquity: "They are of the opinion that things do not reveal the future because they have occurred, but that they occur because they are meant to reveal the future."[54]

From antiquity through the Renaissance, many believed rare meteorological events to be portents. The belief that comets, earthquakes, and rains of frogs or wool presaged plagues, famines, or the death of rulers ran through both poems are not philosophical in their orientation and comment on causation or explanation only in an oblique manner. Melanchthon praised Pontano's *Meteora* for its elegance, which was far removed from what he considered to be the barbaric style of scholastic authors, and for Pontano's dependence on things rather than disputation.[68] Pontano's *Meteora* retained some aspects of Aristotelian meteorology, such as the two exhalations, but its focus was on correspondences between meteorological events and human tragedies. Using the Greek gods as a motif, Pontano linked changes in the weather to wars, famines, overthrown governments, and mysterious destructions of livestock.[69] Explorations of such links, under the encouragement of Melanchthon, became standard for sixteenth-century Protestant meteorological treatises.

Pliny the Elder's *Natural History* also played an influential role in the teaching of meteorology at Wittenberg. Because Melanchthon had written a paraphrase of the prefatory chapter of the *Natural History* (a work that was concerned with the prodigious and marvelous as well as with causation), there were rumors that he had also written the commentary on the second book of the *Natural History*, which was printed under the name of Jakob Milich.[70] While these rumors are probably false, Milich was close to Melanchthon and taught Pliny's work, which discusses astronomy and meteorology, in Wittenberg as early as 1534. Milich believed Pliny's work to be an ideal introduction to these areas of knowledge because of its completeness, and on account of its author's knowledge of ancient sources and Latin.[71] Although Milich found deficiencies in Aristotle's astronomy,[72] he believed that Peripatetic explanations were superior for natural philosophy.[73] Thus his treatment of meteorology retained the terminology of imperfect mixtures and discussed the subject in terms of the four traditional kinds of causation.

Milich's discussion of final causes was cautious although he made a clear attempt to distinguish his views from those of the Epicureans, contending that nature's "ends result from a mind governing all of nature" and echoing the view of Lutheran theologians who equated the denial of providence with Epicureanism.[74] In Milich's view the *physicus* should be concerned with the ends that surround matter, meaning primarily that the *physicus* should explain how "these impressions signify some change in the air or the presence of serene weather or the arrival of a storm."[75] Therefore in his discussion of comets, he explained that comets affect the world through physical causes. They "do not signify so much but take an action."[76] Nevertheless after listing comets whose appearances coincided with historical events, he conceded that observation throughout the ages shows that comets have warned of worrisome events and that "this sign can correctly be considered a final cause."[77] As a result, Milich included signs as part of the final causes of the natural world.

A similar approach to meteorology can be found in the meteorological writings of Marcus Frytsche. Concerns with prodigies and their purpose are paramount in his work, which was first printed in 1555 and reprinted in Wittenberg in 1581, 1583, and 1598. Included with the first and second editions, although not in subsequent ones, was a catalog of strange events or prodigies coupled with a list of major historical events, thereby allowing readers to connect these events to past turning points since the founding of the first city, which he believed happened 1657 years after the creation of the earth. In the sections of his *Meteora* dedicated to the ends of meteorology, he contended that meteorological events are not accidents but that they are signs. They are signs in the sense of being prodigious, both portents of future events and clues to divine providence. For example, he divided rains into two kinds, the prodigious and the natural, just as Pliny had. Prodigious rains are when worms, frogs, fish, milk, hair, rocks, flesh, blood, or iron falls from the sky, as ancient Roman historians and naturalists supposedly witnessed. While they can be explained by natural explanations, such as rising viscous exhalations or the imperfection of parts of earth or mud, these causes are incomplete because "these are fatal and miraculous rains ... that can be called portents."[78] Natural rains, however, are also signs of the divine; they are "rivers of divine providence" that God creates to ensure the survival of plants by moistening the earth "almost like a clepsydra."[79]

For Frytsche, teleology is linked to the larger significance of meteorological phenomena. The final causes of comets are their prodigious meaning. They are portents for "droughts, plagues, famines, wars, and changes in Kings, governments, and laws."[80] Their purpose is bipartite, either pertaining to the temporal needs of humans or reflective of God's particular desires. For example, the rainbow has two final causes; the first, which Frytsche calls physical, allows us to predict rain. The second is theological and signifies that God will not bring another deluge because the rainbow is a sign of the pact between humans and God. More ominously, earthquakes are a sign of God's vindictive nature and of his desire to punish human sins.[81]

Johannes Garcaeus, a professor at Wittenberg, wrote about meteorology in a similar light in his 1584 *Meteorologia*. Arguing against what he considered a Stoic position that there are secondary causes that govern the natural world, Garcaeus emphasized that God's will is the direct cause of meteorology. God's complete freedom means that we cannot have physical causes for all weather phenomena; God is incomprehensible and infinite. Nevertheless, in accordance

CHAPTER THREE

The Ferrarese Earthquakes
and the Employment
of Learned Meteorology

Discussions of the purpose of meteorological phenomena were often inter-twined with religious debates over the character of God's interventions in the natural world and the ability of humans to understand God's will. While these discussions typically took place in universities or seminaries, they often had significant political and social consequences, especially when cataclysmic events affected local populations. Actual catastrophes heightened the intensity of debates, as the grim reality of widespread destruction prompted political and religious rivals to cast blame, encouraged the damaged population to seek remedies, and led scholars to find explanations for disasters. As a result of a pressing need to cope with the destruction of cities, Italian natural philoso-phers, antiquarians, physicians, and church authorities wrote treatises that re-acted to specific disasters, including earthquakes, which were considered to be caused by winds and were therefore considered to be meteorological. These understandings took into account epistemological assessments of the probable nature of meteorology as well as concerns with the purpose and meaning of cataclysms.

During the sixteenth century, the largest number of these treatises re-sponded to the earthquakes that leveled much of Ferrara from 1570 to 1574. The earthquakes that struck Ferrara in these years provoked widespread concerns for safety and caused political and religious turmoil. They also brought out de-bates over the supernatural or natural character of meteorological disasters, reconsiderations of the possibility of apocalypse, and reevaluations of papal au-thority. Writings specifically dedicated to this disaster examined the basis for theological understandings of the seemingly inexplicable destruction. Scholars attempted to determine whether these earthquakes were supernatural by utilizing new observations and by applying accepted theories of earthquakes. The occurrence of earthquakes highlighted the practical applicability of the field of meteorology, as the initial fears of immediate devastation gave way to anxieties over disease and infertility and curiosity over the possibility of predicting these catastrophes.

The thriving genre of disaster treatises that emerged during and immediately after the Ferrarese earthquakes did so in part as a result of conflict between the papacy and Alfonso II d'Este, the duke of Ferrara, and in part because of the duke's need to maintain civil order. Tensions between the papacy and the House of Este had long been strained but were especially tense at the moment the earthquakes began. In the late 1560s, the duke failed to comply with the Inquisition, refused aid to French Catholics, and undermined the pope's crowning of Cosimo I de' Medici as Grand Duke of Tuscany.[1] The tension, however, was heightened as a result of conflicting positions regarding Italy's Jewish population. In 1569 Pius V attempted to ban Jews from all of the Papal States except Rome and Ancona. In the previous several decades, Ferrara had become one of the few centers for Jews in Italy as it accepted Jewish refugees and conversos, who bolstered the local economy.[2] Following the expulsion in 1569, Ferrara received some of the expelled Jewish population, and by 1570 approximately two thousand Jews were living in the city.[3] After the first quake struck in November of 1570, the pope wrote to Alfonso II, contending that the duke and his ministers should have followed his earlier requests to remove the Jews and conversos from the city. The ducal response was that "neither Jews nor Marranos had caused the earthquake, it being a natural thing." The papacy held its ground replying that while the Jews might not have directly caused it, God sent these influences because of the wicked men in the government and because of the conversos, "who are Jews but fake being Christians."[4]

The duke of Ferrara responded to the pope's accusations by encouraging philosophers, physicians, antiquarians, and other intellectuals to write treatises on earthquakes. Their naturalistic interpretations could help absolve him from charges that he was personally responsible for the catastrophe, or at a minimum the interpretations could become a diplomatic tool against the Vatican. In order to undermine Pius V's proclamations that the earthquakes were divine punishment, these authors endeavored to show the implausibility of papal claims without being irreligious. As a result, they used arguments of contemporary Aristotelian natural philosophers. Their ideas about the limited certainty and the possible purposes or meanings of this catastrophe became political tools and a potential balm for the population. Aristotelian meteorology and other versions of learned investigation, while not universally endorsed, were prized by the duke and his circle for their utility to the general public.

ARISTOTLE'S TREATMENT OF EARTHQUAKES

Key to debates over whether the Ferrarese earthquakes stemmed from divine punishment was the evaluation of the conformity (or the lack of it) between these events and accepted Aristotelian explanations and descriptions. Aristotle considered earthquakes, like many other subjects of meteorology, to be difficult to understand and not subject to demonstrative proof. Nevertheless he believed that a coherent theory was possible. He posited the presence of exhalations that circulated underneath the earth, analogous to those above the earth's surface. These exhalations are not only the cause of earthquakes but also the motive force behind the heating of hot springs and the formation of fossils and metals.

Though his exposition of his theory of earthquakes is brief, his justification of it is lengthy. He argued that the earth is dry but contains much moisture. The earth is regularly heated by the motions of the sun and by an internal subterranean fire, which generates internal winds that, at times, flow out from the surface (2.8.365b24–29). Winds, subterranean or not, are formed out of the hot and dry exhalation (2.4.360a18-22; 2.4.361b1–9). The eruption of subterranean winds trapped in the earth's pores or the violent entry of winds beneath the earth's surface are the efficient causes of earthquakes. They are caused by exhalations moving into the earth, into its crevices and hollows, that cause tremors felt on the surface (2.8.365b33–366a5). Aristotle held that the violent nature of earthquakes is evidence that they are caused by the same matter as winds because both are composed of something that is especially strong. According to Aristotle, winds have a prolonged and forceful natural motion, as is clear from the fact that they can move flames extremely quickly.

Aristotle provided significant amounts of corroborating evidence for his theory, even if much of it is vague. He cited just three specific earthquakes yet made broad general claims that depend on or, at least probably should depend on, a broad survey of earthquakes. The confirmations of his theory are categorized as signs (*semeia*). They are signs that the theory is in fact reliable rather than signs that can be used to predict when or where an earthquake might strike, although there is at times some ambiguity in his treatment of earthquakes, as he appears to have taken some evidence from earthquake forecasters.

The signs are as follows. Aristotle believed that most earthquakes occur in calm weather because the exhalation flows altogether in one mass within the earth, leaving the surface free from winds (2.8.366a6–8). Most earthquakes occur at night or noon because these are the times when it is calmest (at night) or when the internal exhalations are strongest (at midday), though he thought that some earthquakes can occur at dawn because this is when the winds usually begin blowing (2.8.366a13–22). The strongest earthquakes happen where the sea is rough or the earth is full of holes because the exhalations are more likely to become constricted underground in these locales. He offered examples: the Hellespont, Achaea, Sicily, and Euboea, all places that were particularly prone to tremors and were near strong marine currents (2.8.366a24–31). Earthquakes are most likely to occur in spring or autumn because the rains and dry spells of these seasons produce more winds both within and above the earth's surface (2.8.366a34–366b2).

That the exhalations cause earthquakes is shown by the specific histories cited by Aristotle. In Heracleia, earthquakes continued until the internal wind erupted from the earth like a hurricane; in Hiera, an Aeolian island, the force of the pent-up exhalations was so great that when it finally escaped from the earth it left nearby Lipara covered in ash. Aristotle believed that it was possible to see the exact point where the exhalation left the ground (2.8.366b31–367a9). Another sign cited in favor of his theory is that the sun necessarily loses its brightness before an earthquake; the wind breaks up and disperses underground. Also, by necessity (*anankaion*), the surface air should be cool and calm during an earthquake as a result of the absence of the exhalation on the surface (2.8.367a21–367b8). Moreover, the exhalation's motion into the earth's bowels causes a thin cloud to form after sunset (2.8.367b8–12). Finally, eclipses are often concomitant with earthquakes because eclipses tend to produce calm weather (2.8.367b20–34). Additional evidence in favor of his theory points to the existence of subterranean winds: they make noises at times, even when they do not cause tremors (2.8.368a14–16).

Aristotle's argument is based on purported observations of regularities that accompanied past earthquakes. He used these observations as signs that his theory is correct. Nevertheless, it is easy to see how at times he verges on giving advice for forecasting, whereby the signs are not signs for the probability of the theory but rather can be taken as predictive signs. If "by necessity" the weather must be cool and calm then noticing unusually cool and calm weather might help predict an earthquake. A thin cloud at sunset, in Aristotle's words, suggests that the exhalations have entered the earth and an earthquake is about to happen. Furthermore, other circumstances—the season, the occurrence of an eclipse, observations about the nature of the ground—might further aid in making predictions. Renaissance Aristotelians systematized these signs, seeing some signs as evidence for theories and others as means by which predictions could be made. Agostino Nifo's commentary illustrates this approach, as he listed twelve predictive signs, including the behavior of birds, tranquility in the air, and the appearance of comets or eclipses. None of the signs are certain, according to Nifo, but nevertheless they assist in prediction. Nifo also allowed for the possibility of astrological prediction of earthquakes, although he deferred to the mathematical experts to determine the exact times and dates.[5]

A number of issues arose surrounding both demonstrative and predictive signs of earthquakes in the aftermath of the 1570 earthquake. Many, if not most, of the conditions established by Aristotle were not satisfied by the events in Ferrara. The earthquakes occurred in the winter in the flatlands of the Po Valley, which are far from violent seas or manifestly porous earth. No coinciding eclipse or thin clouds were observed. The lack of immediately apparent correspondences between accepted interpretations of Aristotle's text and what was observed at Ferrara led some natural philosophers to suppose that the earthquakes did not follow the ordinary course of nature and impelled others to attempt to find ways to reconcile the apparent discrepancies between experience and text. Key to this reconciliation was the belief that meteorological knowledge was provisional and that natural explanations did not rule out other approaches to explaining these events.

DIALOGUES, PROBABLE ARGUMENTS, AND THE CAUSES OF NATURAL DISASTERS

Because many of the writings that addressed the Ferrarese earthquakes of 1570–74 were intended to persuade political elites or the public, their authors adopted modes of communication that differed from the scholarly commentaries of the university. These works are more similar in style and tone to other investigations into natural particulars (such as those on hot springs). These investigations were produced in court settings and differ from the commentaries written by university professors.[6] A large number of these works were dialogues, a genre that was perhaps at its apex in sixteenth-century Italy. In the style of dialogues of the Cinquecento, it is possible to discern their political motivations and courtly audiences. Renaissance dialogues, in general, have been seen as attempts to unite the work of the studio with society at large.[7] Their style was Ciceronian rather than Platonic; elite speakers maintained decorum in realistic settings.[8] Respect and equality among interlocutors prevailed despite the ridicule of Plato's early dialogue and the blind assent to his later ones. Renaissance Italian dialogues generally were pragmatic in nature. Authors adopted the genre in order to increase the market value of the writing, by giving a performance that was entertaining but was nevertheless characterized by tact and a concern for public relations.[9] Thus the dialogue mirrored many of the social and political concerns of the courts.

Vernacular writings on meteorology were relatively common during the Middle Ages and the Renaissance. Meteorology had stood at the forefront of translation movements within Aristotelianism, even during the Middle Ages. In the thirteenth century, Mahieu Le Vilain translated the first three books of Aristotle's *Meteorology* in the 1270s,[10] a century before Nicole Oresme worked on *Ethics, Politics, Economics,* and *De caelo.* During the fourteenth century, an unknown writer produced a work in the Florentine dialect that was a compilation of Thomas Aquinas's and Albertus Magnus's commentaries on the *Meteorology.*[11] In the following century, Evrart de Conty made a partial paraphrase in French of the *Problemata,* a work attributed to Aristotle that treats, among other issues, those connected to weather and winds.[12] Practical utility and the entertainment of the wondrous were motives for rendering the subjects of the *Meteorology* into the vernacular, but they were not the only ones. The field's concern with concrete manifestations of the elements and their mixtures made the subject not only practical but suitable as an introduction to natural philosophy in general. Discussions of prime matter, pure potency, and hylomorphism can confound a beginner as easily as they can enlighten. Some versions of meteorology, however, taught the basics of natural philosophy without reference to such concepts. Thus the elements, the prime qualities, and the relation of the heavens to the sublunary world are revealed by examples found in precipitation, meteors, the seas, and rivers; wondrous examples of meteorological phenomena entertained and gave evidence for God's providence on earth.

During the sixteenth century, a number of university professors and learned courtiers wrote treatises or dialogues on meteorological subjects. Some of these works were meant to bring discourse from universities to a broader courtly audience. The Pisan philosopher Francesco de' Vieri's 1576 *Trattato delle metheore* differs little from the commentaries that were produced from university lectures.[13] The Ragusan philosopher Nicolò Vito de Gozze, on the other hand, transformed university teachings on meteorology into a dialogue that was at least partially intended for a female audience, judging by the preface, written by his wife, and the presence of a female interlocutor.[14] Girolamo Borro, who taught at Pisa, brought meteorology to a courtly audience with even greater élan by writing a dialogue that, using a jocular style, emphasized the marvelous aspects of meteorology.[15] Numerous courtiers, who had not taken up academic careers, also wrote on meteorology in the vernacular. Some, such as Sebastiano Fausto da Longiano, Stefano Breventano, and Cesare Rao wrote treatises; others, including Nicolò Sagri, Camillo Agrippa, and Vitale Zuccolo, wrote dialogues.[16]

The use of dialogue was particularly apt for meteorological discussions because of the inconclusiveness of the subject. Sperone Speroni, the humanist author of analyses of literature, deemed the dialogue to be a form of writing midway between certainty and fiction. He noted that the characters are obliged to speak only in a probable (*probabilmente*) manner about a given subject. In his view, dialogues do not address the certain truth; rather, "The opinion of dialogue is not *scienza*, but a portrait of *scienza*."[17] While Speroni contrasted the alleged certainty of Aristotelian science with the probabilistic reasoning of rhetoric, Aristotle and Aristotelians were aware that the ideals of scientific method were usually not realized in meteorology. For many, it was a subject that could only be treated using probable reasonings. That the unresolved nature of meteorology corresponded to the ambiguous nature of the genre is confirmed by a number of these dialogues, which point to multiple explanations or the contingency of the field's subjects.[18]

It was common for authors to assess the limited degree of certainty that could be attained in this field. For example, Lucio Maggio, in his dialogue on the Ferrarese earthquakes, justified the use of natural principles, found in Aristotle, because Aristotle "has always spoken more probably (*probabilmente*) than others."[19] Giacomo Buoni described his own dialogue as an investigation into uncertain matters according to the varied methods of history, natural philosophy, and theology. Four days of dialogue would be sufficient to "resolve the most difficult matter, even though it is disputed according to probability (*probabilmente*)."[20] Buoni's and Maggio's decisions to write dialogues, in essence, followed Speroni's advice. Speroni wrote, "If I had had certain knowledge (*certa scienza*), I would not have made dialogues, but would have written everything in an Aristotelian [i.e., syllogistic] manner."[21]

Buoni's dialogue on the Ferrarese earthquakes is perhaps the best example of a meteorological text that exploits the dialogue form in its treatment of a contingent and unknowable topic. The inconclusive nature of dialogue allows for the delicate and ambiguous treatment of a politically charged issue. Keeping multiple kinds of explanation available, Buoni emphasized nature's role in producing earthquakes without dismissing divine power. He exploited the dialogue form by making the varied interlocutors illustrate multiple ways of knowing and explaining earthquakes, thereby diplomatically demonstrating the merits of naturalistic explanations while still showing that they could be harmonious with Thomistic understandings of God's power over accidents. By writing a dialogue he could propose both religious and naturalistic explanations, contending that the earthquake was caused by underground winds and that these effects were the result of God's will, which defies any complete understanding. Moreover, his dialogue attacks astrology, a field increasingly being connected to illicit and demonic forms of divination after the Council of Trent. Astrology was also seen as a potential threat to public order because the possibility of subsequent predictions of more cataclysms might foment public hysteria. In this manner Buoni's dialogue entered the political and religious arenas while still being concerned with natural philosophy.

The introductory pages show Buoni's devotion to Alfonso II. Buoni, having dedicated the work to Giovanni Battista Pigna, the ducal secretary and member of the *Riformatori dello studio di Ferrara,* directly praised Alfonso II for his generosity and virtuous spirit. Moreover, the preface affirms the pious nature of this work, stating that "we Christian Catholics know for certain that he [God], not only universally, but also in particulars, has Providence over all things."[22] The early mention of providence signals that Buoni did not dismiss the assertion that God is a cause of the earthquakes, even if he did not accept the papal accusations against Alfonso.

By dividing his *Dialogo* into four days, each being led by a different interlocutor, Buoni was able to present several modes of understanding earthquakes in the voices of different speakers. In the introduction he suggested that the earthquake could be analyzed in multiple ways: "I have determined to treat [the earthquake] not only according to natural things but also by the means of history and theology."[23] In fact we find the earthquakes discussed not just in relation to these three disciplines but also according to medicine as well. He may have omitted that discipline out of modesty, since Buoni himself was a physician. A character based on Benedetto Manzuolo, a philosopher and secretary to the cardinal of Este, led the dialogue on the first day. Alessandro Sardi, a historian, led the second day's dialogue. Sardi later wrote his own treatment of earthquakes that was published in his *Discorsi* in 1587 and dedicated to Buoni.

Buoni led the dialogue on the third day, when medical topics were examined; and Agostino Righini, a Franciscan friar, was the main speaker the fourth and final day, interpreting the destruction at Ferrara from a theological standpoint.

The views of Aristotle; several of his commentators, such as Nifo, Albertus Magnus, and Francesco Vimercati (the last of whom Buoni reported having studied with);[24] and several *novatores*, namely Girolamo Cardano and Georg Agricola, informed the first day's discussion, which centered on natural philosophy. Manzuolo emphasized the limits of natural philosophy, suggesting both that his interpretation is speculative and that references only to nature "are not enough" to fully explain these disasters.[25] Nevertheless, he followed many Aristotelian views of the cause of earthquakes; he contended, "The wind, or spirit, that is closed within the belly of the earth, this spirit, caused by the hot and dry exhalation, whenever the earth is wet, and heated by the sun and the internal fire, immediately makes an earthquake."[26] But earthquakes, while caused by a confluence of natural actions, do not have their own natures; they are accidents of the earth and examples of violent motion because the nature of the earth is to remain still. Earthquakes, thus, are not substantial but accidental and contingent.[27] Even if earthquakes are not substances and therefore are accidental, Manzuolo warned that we should not think of earthquakes as miracles. Chaos exists within the earth, and if parts of the earth move for a limited period, the motion should not be compared with true miracles that affect the entire earth simultaneously, such as the universal deluge from the Old Testament.[28] Thus Manzuolo, emphasizing the speculative nature of his claims, recounted the most common natural explanation for earthquakes without eliminating other potential explanations.

During the second day of the dialogue, Sardi, the historian, describes past earthquakes. Ancient histories and medieval chronicles demonstrate that earthquakes have occurred in numerous locales throughout recorded history, undermining the sentiment that earthquakes are out of the ordinary. Investigations into the past provide more than dates and locales of past tremors; they also show the possibility of new explanations of earthquakes. Sardi's dialogue recounts a story taken from the Byzantine historian Agathias of Mirena, who lived during the rule of Justinian I, about a certain Artemisius, who, wishing to scare his neighbor Zeno after a dispute, made an artificial earthquake that, in the words of Sardi, "imitated nature although not in everything." The contraption worked by copying Aristotle's description of nature, if not nature itself. Artemisius sealed water into various containers and then placed tubes connecting them to an area underneath Zeno's home. Heating the containers with fire, he caused exhalations to rise so that they became trapped, eventually erupting and shaking Zeno's abode.[29] The story, in both Agathias's and Sardi's versions, is meant to entertain, but it also serves another philosophical purpose, that of demonstrating the validity of Aristotle's explanation of earthquakes and suggesting that if a human can devise a way to make the earth tremble, so can nature.

The character Buoni dominated the third day, in which he discussed the medical impact of earthquakes and dismissed the role of astrology. Buoni denied that one can know by astrological means when these disasters will occur. Referring to the ideas of Giovanni Pico della Mirandola and Giovanni Manardi, he bolstered the apparent orthodoxy of his views by lambasting the vanity of the astrologers who asserted that astral causes are determined and constant. According to Buoni, experience shows that astrologers are often incorrect, "and, what matters more, the holy Council condemns them."[30] Following Georg Agricola, Buoni argued that astrologers cannot predict earthquakes and specifically attacked the astrologer Giovanni Maria Fiornovelli, who had made a detailed horoscope of November 16, 1570, which suggested that the prospect of a great earthquake would threaten for some time. Buoni, however, dismissed such a prognostication, which may have fomented public fear in a time when the city was still suffering aftershocks. Buoni wrote that he "had never believed in this vanity that is judicial astrology."[31]

The piety of the interlocutors and their willingness to combine natural and divine causation was a pervasive theme throughout the episode. By the end of the third day, the three interlocutors agree not only among themselves but also with Thomas Aquinas that earthquakes are caused principally by God and secondarily by the eruption of winds closed beneath the earth.[32] On the fourth day, Righini joined them in a new setting, the church of San Francesco in Ferrara. The Franciscan friar dismissed a possible misconception about his order: most Franciscans follow not the Scotist doctors, as might be expected, but rather adhere to the teachings of Thomas Aquinas, whose value to Catholic theology was in the ascendant as a result of the Tridentine councils.[33] Furthermore, the use of Thomas's views would have been a particularly useful tool in undermining Pius V's pronouncements because the pope himself came from a Dominican background, in which his teachers likely would have greatly valued Thomistic theology.

Righini pointed to Frans Titelmans, who used Thomistic ideas regarding the unknowable nature of contingents in his discussion of meteorology.[34] Righini, not surprisingly, stressed the limitations of natural explanations just as

Titelmans had: "And although it will be good, in this consideration of the earthquake, to take care of one matter and not leave behind the other, and to think that the earthquake has its natural causes, one must not stop there, but leap up to God, the author of nature."[35] He concluded the dialogue with a meditation on God's omnipotence: "Returning to the earthquake, it is to be concluded that it is partly natural, and partly divine, that it is sent by God, when he wants, how he wants, where, and how much he wants, and more often for sins, moving with his will the secondary causes, and nature, which he commands at his pleasure."[36] God's will commands nature and his absolute power is beyond human capacity to know, even if it can be surmised that he sends disasters to punish sin. The combination of theology, history, and philosophy is evident in this conclusion, in which the four interlocutors ponder remedies for the destruction. Wells and other holes in the ground might be drilled to let out the subterranean vapors, a tactic used in ancient Rome. A philosophical outlook helps, but most of all it is necessary to pray.[37]

NATURAL AND MIRACULOUS EARTHQUAKES

Politics also shaped Lucio Maggio's dialogue, which was printed in 1571 while earthquakes were still jolting Ferrara. Maggio, a self-described *gentil'huomo*, defended Alfonso II from papal accusations. However, in his dialogue, which was translated into French four years after its initial printing,[38] naturalistic explanations dominate and religion fades away. Just as in portions of Buoni's dialogue, the emphasis on naturalistic explanations was intended to undermine the papal claims of divine punishment. Maggio's primary goal was to demonstrate that the Ferrarese earthquakes were not supernatural. To do this he tried to reconcile the differences between the particularities of the Ferrarese earthquakes and general Aristotelian understandings of the causes of earthquakes, that is, minimize any discrepancies between accepted theories of the natural causes of earthquakes and what was observed at Ferrara, which might suggest that these events were miraculous. Therefore, Maggio adopted Aristotelian categories, using the concepts of imperfect mixture and exhalations as the primary causes of earthquakes, but nevertheless integrated facts about recent events to accommodate his reading of Aristotle with his own observations.

In this dialogue, the interlocutors are all of elite status and again are based on actual people—Fabio Albergati, who later wrote political and moral treatises; a ship-captain, Paolo Casali; the Conte Giulio dalla Porta; and Maggio discuss the causes of the Ferrarese earthquakes. The dialogue attempts to recapture and idealize the learned conversation of the Este court. Captain Paolo, playing the intelligent but unschooled foil for the philosophers, frequently questions how the events at Ferrara fit with Aristotelian doctrine, and Fabio then answers how these apparent contradictions can be reconciled. For example, Paolo says that if earthquakes are caused by exhalations that are heated underneath the earth by the power of the sun, how was it possible for the earthquake to strike Ferrara in the winter when it was covered in ice and snow? Fabio explains that the sun's heat can be stored underground for a lengthy time.[39] Later Paolo suggests that Ferrara was not a likely candidate for the site of an earthquake because it is not cavernous, and Fabio gives an explanation that rules out miraculous intervention as a cause of the earthquake and maintains that earthquakes can occur everywhere. Moreover, Fabio explains that the terrain of Ferrara contains "holes, pores, caverns, and subterranean veins" that are typically filled with water from the Po or recent rains.[40] The heat and dryness of the summer and autumn of 1570, however, dried the holes of their natural humor. Fabio described how he and Maggio in service of Cardinal Paleotti and the prince "together had observed . . . this extraordinary dryness." Because the caverns were deprived of their humors, winds rushed into these holes, because nature does not allow a vacuum, and eventually caused the earthquake.[41] Thus neither was Ferrara a strange locale for an earthquake nor was the event miraculous. Instead these events resulted from unusual but not supernatural meteorological conditions.

In order to bolster his arguments that the Ferrarese earthquakes were natural, Maggio defined miraculous earthquakes. For him, a miraculous tremor was comparable to the earthquake described in the New Testament that reportedly shook the earth when Jesus died.[42] Arguments that this earthquake was miraculous corresponded to sixteenth-century debates surrounding universal floods. Numerous pamphlets and broadsheets, many of which were printed in Germany but some of which were printed in Italy, such as the one written by Tommaso Rangoni, predicted that, because of a grand conjunction in the house of Pisces, a universal or near universal flood would occur on February 1, 1524, at 24 seconds past 3:50 p.m.[43] The argument put forth by these treatises was by no means foreign to Aristotelian thought, which had incorporated astrological causation during the Middle Ages. For example, in the thirteenth century the Dominican scholar Albertus Magnus contended that the cause of the "universal flood" was the conjunction of all seven planets in Aquarius, thereby affirming a natural and celestial basis for this biblical event.[44] This belief in the naturalism of catastrophes was common to other Aristotelian thinkers. Avicenna contended that massive floods periodically destroyed the earth's flora and fauna, the latter of which was renewed through spontaneous generation.[45] Tiberio Russiliano followed Avicenna's general position in his 1518 treatise while adding that according to philosophical arguments, the earth must be eternal and therefore the number of floods and creations is infinite.[46] Russiliano's position was reminiscent of Aristotle's, who not only notoriously argued for the eternity of the universe but also contended in the *Meteorology* that periodic disasters had destroyed civilizations of which we no longer have records or knowledge (1.14.351b9–352a17).

The fearmongering and the potential irreligiosity of those like Rangoni who predicted the flood of 1524 was answered by Nifo, a professor who also had, during his academic career, ties to a number of courts. He attempted to abate the effects of this prediction by attacking its theological and natural philosophical underpinnings. His treatise, which was published in both Latin and Italian, used Aristotelian positions to argue that the planets could not cause such a deluge. Aristotle, in the *Meteorology*, maintained that many of the features of the earth have changed over long periods of time. What was once sea eventually would become land and vice versa (1.14.352a22–25). It could not be the case, however, that the entire earth's composition could change into one element. Rather, there is a natural stasis of the elements and their qualities (hot, cold, wet, and dry) so that while the elements can reciprocally transmute into each other, their inherent balance prevents the possibility of catastrophic global climate change. Nifo used this argument to claim that the predicted flood of 1524 was physically impossible.[47] While Nifo's position might seem to undermine the possibility of universal floods, past or predicted, it instead bolstered religious orthodoxy. If a universal flood was impossible physically, one should not dismiss the story of Noah as false or a fable but rather understand that only divine intervention could have caused it. Aristotle's supposed rejection of the physical possibility of a universal flood thereby supported the position that the biblical flood was in fact a miracle. When the predicted flood did not take place in 1524, Nifo's position gained greater currency, and his argument for the impossibility of a natural universal deluge became common to nearly all commentaries written on Aristotle's *Meteorology*.

Borrowing from discussions that distinguished between the universal flood and particular or local floods, Maggio contended that universal floods must be miraculous, as did most Italian natural philosophers after the affair of 1524. The argument was extended to earthquakes: "The great floods through ordinary causation (*per l'ordinario*) cannot be universal but can only occur in some part of the earth; earthquakes are also the same way."[48] The implication is that universal earthquakes are miraculous and that earthquakes that occur in just one part of the earth are not. Therefore most earthquakes can be explained, at least potentially, by natural principles. For Maggio, his interpretation of scripture supports the past existence of a universal earthquake just as it does a universal flood. In the words of Maggio, "Similarly it was a miracle, that universal earthquake, which shook the entire world during the death of the Redeemer of human kind, Jesus Christ."[49] Ferrara's earthquake, while destructive and long lasting, was small and ordinary in comparison to the truly miraculous earthquakes of the Bible.

Similar naturalistic accounts of earthquakes were found not just in vernacular dialogues but in Latin dialogues as well. Agostino Galesi, who denied that earthquakes have final causes, saw his work as a remedy for the unsophisticated explanations that swamped Northern Italy after the initial earthquake. He proclaimed that "there was nothing more common among literary and inexperienced men than an investigation and dispute over this matter [i.e., the Ferrarese earthquakes]."[50] Galesi, a professor of natural philosophy at Bologna, was acquainted with Maggio and sympathetic to his goals. He rejected earthquakes as portents, contending that political revolutions and the deaths of princes and after earthquakes were just coincidences, and he ridiculed a remedy allegedly prescribed by astrologists of putting effigies of Mercury and Saturn in the four corners of the city's walls. Drilling wells and other small holes in the ground is more efficacious because they allow the exhalations to freely exit from the earth's interior.[51]

Some literary men tried to divorce themselves from Aristotelian positions. For example, the courtier Annibale Romei, in a dialogue printed in 1587 that treats meteorology in general and the causes of earthquakes in particular, offered what he saw as alternatives to traditional explanations. The Ferrarese courtier rejected Aristotle's views on the Milky Way and the cause of the saltiness of the sea.[52] He doubted the warning signs of earthquakes, which were arguably not even maintained by Aristotle, because they do not correspond to examples from history. Nevertheless, despite Romei's proclaimed hostility toward Aristotle, his dialogue covers many familiar themes. General Aristotelian principles, from the dual exhalations to the four causes, form the basis for much of the argumentation. Romei's interlocutors agree with the messages of Buoni's dialogue: astrological forecasts are ineffective, and earthquakes are not miracles but have natural causes. Yet earthquakes, like everything else in the universe, depend on divine will because God governs the earth by using angelic intelligences as heavenly intermediaries.[53]

The Ferrarese earthquakes provoked other witnesses, who were not in service of the House of Este, to discuss the underlying causes. In 1571, Gregorio Zuccolo published a brief Italian treatise on the earthquake and used the opportunity to attack Aristotle. Zuccolo, a scholar of wide-ranging interests and an author of a Latin commentary on the *Analytics*, a treatise on nobility, love, honor, and fortification, and a chronicle of his native Faenza, witnessed the first tremors at Ferrara. He believed his interpretation of this event was based on experience and reasoning rather than on textual authority, thus giving him the opportunity to express his "particular opinions in the matter of meteorology." Zuccolo used the uncertainty of meteorology, a concept endorsed by many university professors, as a rationale for deviating from the philosophy of the schools: Aristotle gives "only a probable argument, not a certain one."[54] Nevertheless, his views appear to be not too far from the Aristotelianism of his day. While he rejected the idea that "windy exhalations" are the efficient causes of earthquakes, he still used terms such as *efficient causes* and argued that for earthquakes the efficient cause is an "essalatione ignibile."[55] Parts of his explanations, however, were more novel. Drawing parallels from the craft of artillery, he contended that the fiery exhalation becomes closed up in pores within the earth. As the exhalation attempts to ascend, it creates heat via friction and eventually catches on fire, causing the earth to expand, which is what happens with powder and artillery, according to Zuccolo.[56] Even though the details of natural explanations were matters of dispute, Zuccolo was in agreement about the incomprehensibility of the phenomena and our inability to understand God's reasons. In fact, these characteristics were common to nearly all of the courtly and academic authors who reacted to the destruction of Ferrara.

ANTIQUARIAN ACCOUNTS

Others who were commissioned by Alfonso II to help counter Papal accusations came from fields beyond natural philosophy. Yet they still could contribute to Alfonso II's general goals, in particular his desire to undermine astrologers' credibility. Pirro Ligorio entered the dispute, unconvinced by Aristotelian natural philosophy yet in agreement with Alfonso II's general plan to restore and maintain political and social order. Originally from Naples, Ligorio, an archeologist and antiquarian, had already had earlier conflicts with Pius V. As an architectural adviser to Pius's predecessor, Paul IV, Ligorio had been a consultant on the use of ancient sculptures in the decoration of his new quarters at the Vatican. Pius V's unhappiness with being surrounded by what he deemed pagan idols thereby cooled Ligorio's professional relationship with the new pope. Having departed Rome, Ligorio arrived at the court of Alfonso II in 1569, where he received a stipend of twenty-five scudi a month and had the opportunity to converse with the poet Torquato Tasso and the historian Alessandro Sardi, among others.[57]

Ligorio's approach to earthquakes was empirical, not only in the sense that it utilized examples taken from the past to gain a better understanding of earthquakes but also in the sense that it rejected theoretical frameworks, at least those based on natural philosophy. Ligorio found that history demonstrates that the Aristotelian position is incorrect. Earthquakes in the past have happened at all times of the day, in all seasons, and in diverse places. Thus the view held by Aristotle and other Peripatetics is contrary to the historical record; according to Ligorio the Peripatetics "want to give rules and a precise period to nature, no more, no less, but what they say is contrary to those things found in history."[58] History shows that "in every age" humans have "attributed [earthquakes] to the secret and interior parts of the miraculous matters of God."[59] If humans have universally found divine agency in earthquakes, there must be some truth in that proposition. This contention is nearly identical to an Aristotelian argument that since all groups of people believe in the gods then the gods must exist. Ligorio used the argument to suggest that the arguments of Peripatetics and astrologers, who, according to Ligorio, deny providence and reduce earthquakes to "defects" of nature, are ineffective because earthquakes are "worthy of reference more to theology" than to natural philosophy.[60]

Ligorio employed theological positions that were in fact not too distant from the Holy See's accusations. In his eyes, earthquakes are "celestial medicine" that God sends down to lead mortals into the divine light and to demonstrate to the impious "how strong and powerful God is." Those worthy of being punished, however, are not the Jews, according to Ligorio, but rather "those who are so bold that . . . they deny God's Providence, having been deceived by Aristotle, Galen, Averroes, Alexander of Aphrodisias, and other Peripatetics." Philosophers are grouped with astrologers as those who "disagree with all good and secure sense" and commit blasphemy when they "take away the lofty power of He, from whom the Greatest Good comes." Rather, for Ligorio, knowledge of earthquakes was limited to the supernatural: God sends earthquakes, and each one is different even in its physical causes.[61]

Ligorio's retreat from naturalism, though it might seem to support the pope, in fact aided Alfonso II. Civil discord threatened Ferrara almost as much as the movement of the earth did in 1570. Astrological predictions stoked the fearful public into an increasingly hysterical state. Ligorio's arguments from history, however, undermined the plausibility of prediction. The ubiquity and seemingly random periodicity of earthquakes meant that they corresponded neither to celestial conjunction nor to natural signs. Therefore the public should ignore the dire forecasts and return to a lawful and pious life. Despite its polemical position toward Aristotle, Ligorio's treatise posed an additional front that partially corresponded to others who wrote on behalf of the Este. Buoni attacked astrology, and both Maggio and Galesi dismissed astrological effigies that were supposed to ward off earthquakes. He believed that these effigies were fraudulent if not dangerous.

In his own treatise Alessandro Sardi, Ligorio's fellow courtier who appeared as a character in Buoni's *Dialogo,* used history to arrive at different conclusions than Ligorio did, even though there were points of agreement. He differentiated between Peripatetics and those who affirm that "everything is made by a divine plan." Sardi thought that those who denied that earthquakes were caused by God were heretics.[62] Writing in his villa outside of Ferrara, where he retreated from the dangers of the city, Sardi emphasized the uncertainty of meteorology and the lack of agreement among natural philosophers: "The cause of the earthquake, variously explained by natural philosophers, was believed uncertain, unknowable, and subject to understanding more through conjecture than truth."[63] Uncertainty about earthquakes was such that the ancient Romans did not even know which god they should propitiate to prevent the earth from shaking.[64] Nevertheless, his treatise is, for the most part, directed toward an examination of the natural causes through historical examples.

Historical analysis provided perspective on the strength and typical duration of earthquakes. The length of earthquakes was of great importance for those experiencing the tremors at Ferrara that continued sporadically for approximately four years. According to Sardi, earthquakes at Bologna in 1504 lasted 136 days, which suggests that the Ferrarese tremors were not without precedent. Additionally, history confirmed the conditions and locales that lead to earthquakes, suggesting, in Sardi's view, a naturalistic interpretation of the recent catastrophe. While admitting that God "rules, disposes, moves, and halts natural causes," Sardi thought that there was also a natural reason that Ferrara was subject to earthquakes, a reason that might be hidden to many casual observers or to those unaware of Ferrara's past. Sardi, who claimed he wrote a treatise on the ancient topography of Ferrara, maintained that in ancient times the territory around Ferrara had been reduced to a swamp. Then the Po River washed mud onto the site of Ferrara, raising it from its former marshy altitude. As a result, the earth immediately beneath the city appeared not to have many caverns into which the exhalations could flow. Deep below the surface, however, there were "great caverns" that were the locations for "making a powerful and lasting earthquake." Changes in the terrain over time are the reason that in antiquity the earthquakes in the Po Valley were weak. By the Middle Ages, however, the influx of mud had rendered Ferrara susceptible to tremors. The accuracy of Sardi's geological chronology is corroborated by the earthquake that shook Ferrara on December 13, 1285, which was "assai robusto," not like the delicate ones of antiquity.[65]

Sardi's historical analysis, while at times pessimistic, retained some hope for his suffering contemporaries. The Greeks and Romans who thought that earthquakes indicated future evils were justified. According to Sardi, plagues followed earthquakes in Rome in 316 AUC (437 BC), in Constantinople in 801, in Northern Italy in 1348, in Modena in 1501, and in Ferrara in 1505. Examples of coinciding famines and wars suggested the ominous nature of earthquakes, which can be explained by demons, as Porphyry had explained, or, when "speaking naturally," they could be explained by the corrupting air that rushes from the bowels of the earth. Historical investigation, however, also gave clues to divine providence, according to Sardi. The earthquake that struck Constantinople in the thirtieth year of the reign of Theodosius was divinely announced to the emperor and the patriarch Procolus.[66] This divine message revealed that God through his kindness "punishes, but does not destroy his sons." Thus history demonstrates that prayer, according to Sardi, is efficacious and that true Christians need not fear total destruction.[67]

The physicians, natural philosophers, and antiquarians who aided Alfonso II employed various methods and came to vastly different conclusions. Some aligned themselves with Aristotle and interpreted recent events in light of ancient texts. Others used ancient texts to demonstrate the implausibility of natural philosophy and the impossibility of prediction. All, however, admitted that the role of the divine in causing particular earthquakes was unknowable. Maggio determined that the Ferrarese earthquakes were not miraculous, despite identifying and describing the class of miraculous universal tremors. Buoni utilized Thomas Aquinas's views on knowledge of contingents to chip away at the assertions of a pope who came from the Dominican order. Just as Buoni considered multiple kinds of causation, Sardi distinguished demonic explanations from those based entirely on natural causes by appealing to his understanding of history and the changing topography of the Po Valley. Ligorio believed God to be the direct cause of earthquakes, but his conviction that earthquakes could occur anywhere aided those who maintained that the event did not violate reasonable possibilities that earthquakes could occur in the Po Valley. The multiplicity of explanations reflected the uncertain status of meteorology and, in particular, of knowledge that related to earthquakes. Establishing uncertainty was paramount to Alfonso II in his diplomacy and civic goals. Uncertainty about causation rendered forecasts unreliable, thereby helping to restore civic order. More important, uncertainty about how and why, and even if God causes earthquakes weakened the pope's position and reasserted the piety of the House of Este.

More than a century after the Ferrarese earthquakes, Pierre Bayle attacked those who interpreted comets as supernatural signs. Part of his critique was directed at ancient and medieval historians who too readily included in their works any reports of miracles and prodigies to please the reader and make their works seem more poetic. For the skeptical Bayle, the inclusion of reports of miracles made histories unreliable and led scholars, like Frytsche and Meurer, to conclude falsely that comets were prodigies and harbingers of catastrophes. Accurate investigations into the past, according to Bayle, reveal that there were no more disasters in years of comets than in other years.[68]

Ligorio's and Sardi's appeals to ancient history differ from those of Lutherans such as Frytsche and Meurer, who used chronicles to support their contention that meteorological phenomena were signs of God's will. Ligorio thought that records from the past undermined any attempt to find precise rules about the occurrences of earthquakes. Sardi, while still open to the possibility of supernatural causation, saw history as a way to investigate naturalistic explanations. History could provide evidence of the changing terrains of specific areas that are the material causes of earthquakes. Sardi's investigation into the past also provided him with perspective. Descriptions of past earthquakes suggested that the tremors that Ferrara experienced had parallels in the past and therefore might be part of the natural order. For Sardi, as for Bayle, history aided natural philosophy.

Firsthand investigations into the material causes of earthquakes led to reconsiderations of the circumstances that surrounded the Ferrarese earthquakes and led natural philosophers toward new approaches. Maggio critiqued and modified what he thought were accepted explanations of earthquakes because they did not correspond with what he observed in Ferrara. Zuccolo and Buoni considered siege mines and artificial constructions to be models for understanding earthquakes. Buoni looked to Byzantine history to find his inspiration, while Zuccolo looked to emerging technologies. Both, however, saw that the artificial could explain the natural. Ancient texts and histories remained present in later meteorological works; but Zuccolo's use of gunpowder and artillery to explain earthquakes resonated with a number of late sixteenth-century meteorological works, as some natural philosophers modeled explanations of meteorological phenomena on artificial explosions and attributed the material cause of lightning and other aerial fires to chymical substances. As the next chapter will show, the use of chymical explanations changed the nature of Aristotelian meteorology during the years that followed the Ferrarese earthquakes.

The Chymistry of Weather

Early on the Saturday morning of October 6, 1646, a purple-colored rain began to fall on Brussels. The amount was great enough that it flowed through all the city's rivers and canals. To the astonished locals its hue recalled wine or even blood. The fame of this rain spread throughout the region, bringing a flow of curious onlookers to a Capuchin monastery, where monks had captured a large sample of this liquid in a barrel. Experts came, forming a committee of church officials, some of whom tasted the mysterious precipitation. Many witnesses undoubtedly considered the rain to be miraculous and prodigious, as a number of sixteenth-century meteorological texts had discussed bloody rain as ominous and portentous.[1] The response to this event in 1646, however, was distinct from many earlier discussions of rain of this color and from the politically charged debates over the causation of the earthquakes that occurred in the 1570s. Rather, a group of "famous men" downplayed any theological significance behind the event and attempted to give explanations of the rain's color largely on the basis of chymistry. Their letters were then collected and first printed in 1647.[2]

The experts distinguished the rain of 1646 from an earlier one that had fallen on much of Western Europe in 1608. The purple rain of 1608, they concluded, resulted from the excrement of an enormous migrating rabble of butterflies. The experts concurred that the more recent rainfall could not be attributed to a similar cause. And while explanations varied, they remained fixed on the idea of aerial chymical reactions. Govaart Wendelen, a canon, cited theories that explained extreme kinds of lightning by the presence of niter, sulfur, camphor, and charcoal (107). Jean-Jacques Chifflet referred to bituminous vapors that contained asphalt and chalcanthum, a copper sulfate solution that was a common ingredient in inks (110–11). Responding to Pierre Gassendi's doubts, Chifflet argued that he had conducted experiments that confirmed his position (120). Even though Gassendi and Vopiscus Fortunatus Plempius questioned the presence of chalcanthum, Chifflet's positions and techniques were by no means far from standard.

Allen G. Debus has demonstrated that Joseph du Chesne and other adherents to the Paracelsian tradition applied chymical concepts in their explanations of aerial phenomena.[3] They were not alone. By the late sixteenth century, some Aristotelian natural philosophers and their opponents had transformed the matter of Aristotle's dual exhalations from imperfect elemental mixtures to combinations of chymicals. Some even used techniques such as distillation to bolster their arguments. Even though Paracelsians raised awareness of chymistry, a tradition of applying chymistry to natural processes arose within Aristotelian thought. Many strands of late Aristotelianism were extremely eclectic and incorporated novel approaches. Renaissance natural philosophers turned to a variety of sources to find evidence for new conceptualizations of the two exhalations. These new conceptualizations were crucial to developments in late Renaissance meteorology because understanding the underlying matter of meteorology was one of the primary goals of those working in this field, in which material causation was far more important than formal and final causation, at least according to Aristotle.[4]

Renaissance natural philosophers aspired to form theories about meteorology that corresponded to experience, that were at least potentially supported by textual authority, and that did not have ramifications that led to impossibilities. One major difficulty was observing the phenomena. Lightning and earthquakes are transient and occur in remote locations, so they could seldom if ever be the subject of lengthy visual analysis. Conjectures about the material substrate of various fires in the sky depended on analogies and comparisons to phenomena more accessible to observation and on the few physical traces that these fires left on the surface of the earth. Phenomena occurring on or just below earth's surface provided potential explanations for what takes place in the upper regions of the sky. Aristotle had given license to apply explanations of nearby phenomena to those phenomena that are far away when he concluded that "the same natural substance causes wind on the earth's surface, earthquakes below, and thunder above."[5] That substance was the terrestrial exhalation. The unity of Aristotle's meteorological theories thereby indicated a method.

Explanations of accessible and observable meteorological phenomena could be applied as clues and analogies to elucidate the remote.

Many natural philosophers of the sixteenth and seventeenth centuries followed Aristotle in at least portions of this statement, even if they did not consider themselves Aristotelians. Renaissance opponents of Aristotle, such as Giovan Battista della Porta, had no better method to observe the upper parts of the sublunary region than their predecessors and thus relied on analogy, using explanations for the subterranean phenomena to explain the aerial regions. Nevertheless, even though the employment of analogy continued, as did the reliance on the idea of the hot and dry, or terrestrial, exhalation throughout the seventeenth century, ideas about what this exhalation was and explanations for its properties changed significantly during the sixteenth century. In particular new conceptions of the composition of the subterranean regions altered explanations for lightning, thunder, and other phenomena of the upper air.

The biggest change during the Renaissance in the understanding of the exhalations was the increasing reliance on chymical explanations and methods. Lightning was attributed to the explosion of sulfur in the air; earthquakes and volcanoes resulted from bituminous soils; and niter coursed through the winds. Natural philosophers saw the most basic chymical procedures—distillation and precipitation—as analogous to the natural rising of vapors from the earth and the leaving behind of residues; and more complex chymical procedures, such as the reduction to the pristine state, were used to analyze the composition of meteorological substances. Paracelsus had some influence in the growing dependence on such chymical explanations, by which he himself described the causes of weather. However, the motivations behind many of the adoptions of these kinds of explanations came from new interpretations of Aristotle's exhalations that integrated his view with the authoritative writings of Seneca and Pliny, firsthand observations, chymical investigations, the modeling of meteorological events on bombs and firearms, the findings of late medieval balneology, and Renaissance writings on metals and mining, including writings by Georg Agricola and Vannoccio Biringuccio. These varied sources contributed to the preeminence of sulfur, bitumen, and related unctuous substances in Renaissance meteorological explanation.

ARISTOTLE'S *METEOROLOGY* AND THE SOURCES OF FIRE

Aristotle left the precise nature of two exhalations undefined to a certain degree. This allowed later natural philosophers to interpret his theory as chymical and corpuscular, as it is possible to reconcile his descriptions of the exhalations with chymical and mineralogical knowledge. Especially difficult for Renaissance thinkers was understanding what was the source of fire below the earth and in the sky. The *Meteorology* does not strictly follow the outlines found in other works. The presence of a fiery substance in the upper regions of the atmosphere fits with the general outlines of Aristotelian physics because fire is naturally light and so its natural motion is upward. His analysis, however, of what is below the surface of the earth does not correspond with the elemental earth's primary qualities: dryness and coldness. The earth, as a whole, is not just elemental earth but a mixture of all elements, and it has an internal source of heat. Aristotle's pronouncements on this subject are vague, and it is unclear why the internal source of heat does not exhaust itself and what its material constituents are. In his discussion of earthquakes, he referred to an "internal fire" that heats fallen rain, thereby creating underground winds (*pneuma*; 2.8.365b26). Aristotle mentioned the earth's internal heat several other times in the *Meteorology*, though without much illumination about its causes or material source. Nevertheless, a large amount of fire and heat within the earth is presented as the cause of underground exhalations (2.4.360a6, 2.4.360b32). The subterranean presence of the hot and dry exhalation is crucial to the formation of fossils, as a material cause, and metals, as an efficient cause.[6] A brief discussion of the taste of spring water suggests what might be one of the causes of this heat. Aristotle believed that salt rivers and springs were once hot, but the fiery principle (*arche puros*) had been extinguished, although the earth through which the water is filtered has retained alum (*stupteria*) and ash (*konia*), which have retained the quality of fire (*dunamis puros*; 2.3.359b5–21).

While it is unclear how these salt rivers and springs originally gained their principle of fire, a potential solution emerges from the *Problemata*.[7] In problem 18 of *particula* 24, the author asks, "Why are warm bodies of water all salty?" The first potential answer roughly corresponds to the *Meteorology*. The earth surrounding springs often contains alum that filters the water. The second answer asserts that lightning bolts have struck these bodies of water, leaving behind sulfuric ashes. The author repeated the second answer in the next problem, in which he asks, "Why are hot springs considered sacred?" Possibly, it is because they have their origin in the most sacred things: lightning and sulfur (24.18.937b21–28). That the residue of lightning is the cause of the hot springs is an attractive solution because Aristotle ruled out the possibility of a unidirectional cooling or drying of the earth. The sublunary world is an eternally stable closed system. While a particular hot spring might eventually cool, another one could arise elsewhere after being struck by lightning. But perhaps what is more relevant is the suggestion that the exhalation that causes lightning is composed of sulfur and that understanding the composition of matter on the earth's surface might have some relevance to the material composition of fires in the sky.

While the matter of lightning (i.e., that subspecies of the dry exhalation) might be considered sulfuric in the *Problemata*, Aristotle's descriptions of this exhalation and fiery meteorological substances are still fairly vague in the rest of the Aristotelian corpus. In *Meteorology* 4.9, Aristotle defined the secondary passive qualities of homeomerous substances, adding a bit more detail about the material basis for inflammability. Inflammable bodies (*ta phlogista*) are not moist yet contain fumes (*ta thumiata*) and produce smoke. Pitch (*pitta*), oil (*elaion*), and wax (*keros*) are more inflammable when mixed with other substances, presumably because of their moistness (4.9.387b18–24). The fattiness of pitch, oil, and wax allow them to burn for long periods of time. Nevertheless, the lack of detail of Aristotle's description of inflammable bodies, his obscurity, and his recognition of a number of subspecies of the dry exhalation made it easy to accommodate a diversity of phenomena under this concept. At first glance the differences between *pneuma*, smoke, sulfur, and fattiness would seem to prevent a unified understanding of the hot and dry exhalation. But once having been made, the links among *pneuma*, inflammability, and the oily and unctuous enjoyed a long afterlife.

THE STOICS' NATURAL PHILOSOPHY

Although Aristotle used the term *pneuma* for the subspecies of the dry exhalation that causes winds, the term had a different meaning and a more central place in Stoic thought. For Stoics, *pneuma* was a combination of fiery and aerial principles that interpenetrated all matter, rendering the universe dynamic and stable through the principles of contraction and expansion.[8] The cross-fertilization of Aristotelian and Stoic natural philosophy during the Middle Ages led to the prominence of the concept of unctuous moisture. Unctuous moisture was believed to be a fatty substance that interpenetrated substances, in the same fashion of the Stoics' *pneuma*, and thereby gave cohesion and transtemporal stability to otherwise earthy materials.[9] The presence of unctuous moisture was particularly strong in substances such as bitumen and sulfur that were considered to be highly inflammable and were identified with the dry exhalation.

In a sympathetic rendering of Aristotle's theory of the cause of lightning, Seneca emphasized a corpuscular explanation while using sulfur's readiness to burst into flames as an analogy to events in the sky. Seneca described Aristotle's exhalations as "great supplies of corpuscles (*corpusculorum*) which the earth ejects and pushes into the upper region." Some of these exhalations that emanate from earth, that is, hot and dry exhalations, go into clouds where they are the fuel for fires, being ignited either by collisions or by the sun's rays. Seneca likened these particles in the upper air to what we see on earth, when matches sprinkled with sulfur "attract fire from a distance." (*Naturales quaestiones*, 1.1.6-10). Thus the corpuscular explanation is intertwined with one that asserts that particular substances, such as sulfur, are potentially fiery and that their particles are causes of fire, both on the earth's surface and in the upper air.

Experience attested to the fiery nature of sulfur and similar substances. In Seneca's *Naturales quaestiones*, lightning and some spring waters are naturally sulfuric. As a result, the former causes disease and the latter both causes and cures them. Lightning condenses and congeals liquids that it strikes so that there are more links (vincula) between the corpuscles. The presence of numerous links is the cause of viscosity. Seneca believed that lightning is fiery because of the supposed fact that everything it strikes smells of sulfur (2.20, 3). The fiery particles of lightning have a sickening power and leave behind the odor of sulfur, which, according to Seneca, causes madness if inhaled (2.53). Similarly, in his discussion of the properties of water, Seneca described medicinal waters as being affected by sulfur, iron, or alum (3.2, 2). These substances potentially can be dangerous. The differing temperatures, tastes, odors, and exhalations that emanate from springs depend on whether the water has been filtered through earth that is full of "sulfur, niter, or bitumen" (3.20, 2). He explained that water assumes the heat and flavor of sulfur or niter as it passes through it. Evidence that materials similar to sulfur can alter water is found in "quicklime" that, according to Seneca, will boil water that is poured on top of it (3.24, 4).

The fixation on sulfur was limited because Seneca used multiple explanations for the causes of hot springs. He attributed one to Empedocles that employed analogies to artificial constructions. Seneca claimed that Empedocles believed that below the earth there are many subterranean fires that heat springs. These fires heat water that passes through underground passages that are similar to *dracones* and *miliaria*. These devices were pipes that coiled over a fire so that the descending water would spend more time close to the fire. Seneca believed that a similar process happened naturally at Baiae, near the Bay of Naples, where the baths were heated without recourse to human-made fires—cold water flowed through passages that were heated naturally by subterranean fires (3.24, 2-3). The analogies between the natural and the artificial as well as the recourse to specific kinds of flammable substances influenced later meteorological and subterranean explanations.

Seneca was not the only ancient Roman author to proffer mineralogical explanations for subterranean heat. Pliny the Elder's account of meteorology in the second book of the *Natural History* partially follows Aristotle's explanations; for example, the two exhalations play a large role. Although Pliny did not elaborate on the material composition of the exhalations beyond the fact that one was terrestrial and the other humid, his accounts of the subterranean describe the fiery substrate that creates hot springs and volcanoes. His description of subterranean fires is significant for its description of the matter of fire (*materiae ignium*; 2.239). Although he did not specifically affirm that this matter was also the substrate for aerial fires, this conclusion was implied. If some kinds of earth are more likely to ignite and fuel fires, and if the terrestrial exhalation is composed of particles of earth and minerals, then it is reasonable that these fiery particles cause the fiery phenomena of the atmosphere when they are lifted into the air. Pliny described a small number of marvelous fires, all of which have a similar substrate. The first example he gave was a swamp in Samosata that emits lime (*limum*) that burns (2.235). The Samosatians used it to protect their walls from the invading Lucullus. According to Pliny, a substance called *naphtha* is similar to this burning lime. Naphtha, a word still used to describe products from distilled petroleum, was, according to Pliny, found in Babylonia and Parthia and burns in a similar way to what he called liquid bitumen, which for him is the apparent material cause of a number of volcanoes and naturally hot earthly craters (2.235–38). Pliny's and Seneca's accounts appear to have influenced both medieval and Renaissance understandings of the material substrate of fire and lived on into the seventeenth century when sulfur, bitumen, and similar substances were believed to cause aerial meteorological phenomena.

MEDIEVAL MINERALOGY

The portion of Avicenna's *Kitāb al-Shifā'* that came to the Latin West under the title *De congelatione et conglutinatione lapidum* influenced debates on the efficacy of chymistry as well as theories about the formation of geological bodies. Avicenna contended that stones were made out of "viscous mud" (*viscosum lutum*) that was transformed by a mineral force or power. The viscous material has affinities to the oily vapors that are the matter of fire. In this work Avicenna provided more evidence for the identity of the matter for subterranean and sublunary formations. One piece of evidence is found in the formation of lightning stones. These stones, later classified as *ceraunia* and now considered to be arrowheads or other prehistoric artifacts, were attested to by Pliny the Elder, who believed they were formed from the residue of lighting.[10] Avicenna explained that these stones were formed in the same manner as certain colorful terrestrial stones: "Often also stones are created out of fire when it is extinguished, since often earthy and stony bodies fall during thunder, [and] because fire becomes cold and dry by its extinction. And in Persia during thunder, aerial bodies fall and things similar to hooked arrows that cannot liquefy but by fire evaporate into smoke as their humidity is thickened, until the residue becomes ash."[11] Thus stones can be formed in the air or underground if there is sufficient fire and the presence of the appropriate material to be burned.

That this material fit for burning is unctuous moisture became more explicit in Albertus Magnus's writings. Albertus helped codify the idea that there are two kinds of moisture: one that is aqueous and the other that is unctuous. The unctuous moisture is akin to the radical moisture that is the life-preserving substance of living things as well as the underlying cause of the vital heat of animals. While the unctuous moisture is difficult to separate from the bodies that contain it, the aqueous moisture is easily vaporized. Albertus explained the differing properties of these two kinds of moisture by appealing to the process of distillation and the strong connections of the particles of oily substances that "are connected like the links of a chain and cannot easily be torn apart." Moreover, unctuous moisture is "easily inflammable and is active in burning things with which it is joined."[12] Unctuous moisture is in some minerals the material principle of inflammability, a principle common to sulfur, bitumen, and naphtha, according to Albertus. In his discussion of alums, which are naturally formed hydrous sulfates, he linked alums to the residues of petroleum, namely bitumen and naphtha: "It is said that there is found a 'moist alum' and that it is like unctuous bitumen, very easily consumed by fire; and in this property and in being unctuous, it is very like sulfur, but lacks its odor. And this form of alum some people call naphtha."[13] Thus for Albertus, sulfur and its wet fattiness are the models for the substrate of the inflammable.

Albertus applied this model of inflammability in his commentary on Aristotle's *Meteorology*, a work that was considered the most authoritative medieval commentary on this subject during the Renaissance.[14] In accordance with his conviction about the presence of sulfur underground, Albertus attributed the cause of some earthquakes to the interaction of wind and underground fires, using analogies from the earth's surface: "For we see that wind shakes everything and moves still bodies on the earth's surface. For it moves fire, sometimes blowing it away from combustibles and extinguishing burning bodies; but sometimes by inflaming burning bodies and by making the fire penetrate them so that the interior seethes and burns up. Sometimes [wind] becomes lit by its own agitation, just as in a lightning cloud, sometimes entering into the pores of the earth it starts a fire, by means of friction, in the sulfuric and arsenic matter."[15] Thus Albertus concluded that the motions of the hot and dry exhalation, which is the matter of wind as well, are so powerful that they are capable of causing tidal waves, whirlwinds, floods, and the collapse of "buildings, big structures, trees, and huge rocks."[16] Perhaps more pertinent is Albertus's proclamation of a unified material cause, the terrestrial exhalation, for wind, earthquakes, and lightning.

Albertus's discussion of lightning gives a more precise understanding of the nature of the hot and dry exhalation. For one, there are variations in exhalations, since the terrestrial exhalation is mixed in different proportions with aqueous vapor, which produces corresponding degrees of denseness. The array of colors of lightning results from the exhalation's tenuousness (*tenuitas*), compactness (*spissitudo*), closeness (*glutinatio*), or viscosity (*humidum viscosum*).[17] Albertus repeated, seemingly with approval, Seneca's contentions of the sulfuric nature of lighting. Although he found Seneca's words not entirely adequate, he agreed that objects struck by lightning have a sulfuric odor and are fetid "because the wateriness of the vapor [of the cloud] is mixed with a terrestrial heat, that corrupts, burns, and brings about a sulfuric unctuousness. And such a vapor moves from watery clouds having burning terrestrial vapors within itself."[18] A similar process brings about the generation of sulfur and causes all things struck by lightning to seem sulfuric. While Albertus did not strictly identify the hot and dry exhalation with sulfur, he did believe there were similarities. He saw a commonality in the presence of the viscous humor and found that sulfur could be the matter of fire both above and below the ground.

BALNEOLOGICAL INVESTIGATIONS

The importance of sulfur in the study of the natural world grew during the later Middle Ages not just because of its fundamental role in mineralogy but also through a growing interest in natural baths. Just as Seneca's and Pliny's ideas about the presence of subterranean sulfur was formed in part by their familiarity with naturally warm waters that smelled like sulfur, medieval physicians, particularly Italian ones, associated the curative properties of natural baths with the powers of sulfur and other inflammable minerals. Growing interest in the medicinal powers of bathing during the fourteenth and fifteenth centuries led to a corresponding familiarity with a large number of baths, many of which had been recently discovered or rediscovered. Economics motivated further investigation of the properties of these baths as they became sources of income for municipalities, patrons, and proprietors.[19]

A variety of individuals wrote on this topic, some of whom had few ties to the university or the teachings of Aristotle. In the Aristotelian schema, however, the causes behind these springs' heat were part of meteorology; they were relevant to the first chapters of the second book of *Meteorology*. Thus in the last decades of the fifteenth century, the extremely erudite Paduan physician Michele Savonarola combined his knowledge of natural philosophy and medical theory with his own examination of these baths and springs as well as with considerations of results taken from chymistry. Over a century before Savonarola, Iacopo Dondi had pioneered using distillation and the analysis of exhalations, which Savonarola considered along with his own experiences in his investigation of baths. Following the stance of prominent writers on meteorology and a major line of medical thought, Savonarola considered his conclusions probable and experience paramount in the investigation of springs.[20] This probabilism did not lead him toward a rejection of natural philosophy—the Aristotelian concept of antiperistasis remained a tool in the causal explanations of the temperature of water in caves.[21] Nevertheless, his genuine emphasis on experience led him to consider the compositions of numerous thermal baths in Italy based primarily on his own observations.

Savonarola's experiences led him to conclude that bitumen played a dominant role in the formation of hot springs. He hypothesized that some underground hollows are full of bitumen, a material that, in his view, is particularly unctuous, solid, and compact—one that burns continually because of the power of the sun and the stars. A high concentration of this bitumen creates fervid water that cools while traveling through the earth's pores and eventually arrives at the earth's surface.[22] Even though he thought that bitumen was more important than other minerals in the production of underground heat he noted that other minerals were key to forming the curative properties of different baths; for example, that sulfur predominated at the baths of Petriolo and that the baths at Abano were full of alum helped explain their particular healing powers.[23] Savonarola, concerned with the differences that gave each bathing site its particular characteristics, engaged in more general descriptions of the minerals that were the fuel for the heat of these springs. Alum, sulfur, and bitumen, for Savonarola, were remarkably similar; wet alum is similar to bitumen in its unctuousness and inflammability, although it lacks the smell of sulfur.[24] And although bitumen is the source of heat for water, it is sulfur that is responsible for volcanic eruptions.[25] Whether or not his particular evaluations were broadly accepted, the experiences and observations of Savonarola and other authors of balneological tracts gave further credence to the importance of fatty inflammable substances such as bitumen and sulfur in the production of subterranean heat.

RENAISSANCE MINERALOGY

Savonarola and other physicians, while trained in Aristotelian natural philosophy, distanced their investigations into the nature of hot springs from traditional university teachings. Similar emphases on practical fields of knowledge, personal observation, and the minimization of university teachings are found in the works of several early sixteenth-century investigators of minerals, Vannoccio Biringuccio and Georg Agricola, both of whom encouraged chymical analyses of the natural sources of underground fires. Biringuccio gained most of his knowledge of nature from his engineering projects rather than through the study of texts. His examinations of pyrotechnics, a field that considered artificial fires writ large, influenced Renaissance meteorology through his comparison of earthquakes to the effects of the recently invented siege mine. Such a comparison would be long lived, as later meteorological writers would move the analogies from under the earth to above, trying to find similarities between lightning and gunshot. Furthermore, Biringuccio's analysis promoted a chymical understanding of earthquakes because his recipe for siege mines called for placing compact quantities of acid, sulfur, and sal niter in hollows beneath the ground. When ignited, the mixtures would impel exhalations to move the earth above. Even though Biringuccio did not argue explicitly that earthquakes occurred by the ignition of the same material, his comparison of their effects would suggest to later writers that this might be the case.[26] For example, appealing to the military sciences was a mode in which Gregorio Zuccolo could dismiss the idea that the windy exhalation was the efficient cause of earthquakes, as he argued that their suddenness paralleled the explosions of artillery.[27]

The mineralogical writings of Georg Agricola also had a significant impact on considerations of the causes of subterranean fires and their chymical composition. Agricola maintained in the traditional vein that exhalations enclosed underground were responsible for seismic phenomena but divorced his view from what he considered to be Aristotle's explanation. Using evidence taken from his own visits to mines, where he determined that the heat generated by the sun cannot penetrate very far beneath the earth's surface, he rejected the contention that the sun could heat these exhalations. This view was also supported by ancient authorities; Theophrastus in his *De causis plantarum* proclaimed that the inability of the sun's rays to heat deep below ground level was proved by the fact that plants' roots do not extend below a certain level.[28] Agricola allied himself with another later Peripatetic, namely Strato of Lampsacus, in his explanation of earthquakes, when he argued that after the subterranean fire has altered vapor, the cold surrounds it.[29] Because heat and cold cannot coexist in one location, the vapor expels itself, thereby shaking the earth. Presenting himself as opposed to Aristotle and to those he identified as astrologers, Agricola contended that the heated exhalations that cause earthquakes must gain their heat not from the sun but by an internal fire, a view that in fact is not necessarily foreign or contradictory to Aristotle.[30] Nevertheless, the existence of an internal fire raised a potential problem. What prevented the fire from consuming itself and becoming extinguished? Earlier solutions, endorsed by Albertus and Savonarola, relied on the heat of celestial bodies to renew this heat, a view that Agricola dismissed. Agricola's solution relied on his knowledge of the various kinds of minerals that resulted from his knowledge of classical texts and his experience with mining.

In his *De ortu & causis subterraneorum*, Agricola wrote that there were two possible candidates for the matter of earth's internal fire: "Of all of those things that are dug out from the ground, there is nothing that produces fire more easily than bitumen and sulfur, on account of their fattiness."[31] Of these two possibilities, Agricola eliminated sulfur because it burns quickly and is extinguished by water, properties that suggest it cannot be the cause of a perpetual fire or of hot springs. Bitumen, in contrast, burns in the presence of water, which even enlarges the flame. That bitumen reacts in such a way when exposed to water might be considered additional evidence that it heats hot springs. While bitumen's powers may last longer than sulfur, both are present in the earth: bitumen is "a fatty juice similar to oil," while sulfur is "the fat of the earth, from which the force of heat is expelled." Agricola appealed to the fact that sulfur is found in great quantities in burning places, presumably places such as volcanoes, and that a sulfuric residue is left in the crucible when pyrite and flaky stones are liquefied in laboratory testing.[32] For Agricola, however, these two substances are similar, as are a range of other fire-inducing minerals. Demonstrating his lexical abilities, Agricola wrote in his *De natura fossilium*, "Another fatty juice follows, linked by natural understanding with sulfur, which the Greeks call *asphalte,* the Latins name bitumen." These minerals, slightly different in their qualities and in the locations where they are found, form a large group of minerals including "naphtha, camphor, *maltha, pissasphaltus, gagates,* Samthracian gem, Thracian stone, *obsidianus lapis,* and many others listed by Pliny."[33]

Agricola's favoring of bitumen over sulfur as the source of the earth's internal heat did not convince all. Andrea Bacci, a physician, an antiquarian, and the author of an influential and authoritative work on balneology, which was first published in 1571, disagreed with Agricola. Bacci contended that the study of baths was crucial to both medicine and natural philosophy and thus included discussions of the nature of various kinds of bodies of water and types of earths. Although he did not mention Agricola by name, Bacci referred to a recent view that favored bitumen over sulfur as the "opinion of writers from Germany." Rejecting the idea that bituminous fires are perpetual, he cited Mons Gibium (present-day Salvarola) near Modena where the well water is distilled with bitumen but nevertheless remains "cold, even though its smell is pungent and its hue is purple." He made further appeals to the sense of smell by arguing that the odor of bitumen never permeates the sites of hot springs, while the stench of sulfur is always present. Additional evidence arose from the realm of anthropology, as Bacci purported that it is the consensus of all nations that the sacredness of bathing can be attributed to sulfur. He cited Pliny, the *Problemata,* and the fact that the Greek word for sulfur, *theos,* was also their word for the divine.[34]

The growing interest in Pliny during the middle of the sixteenth century provided an additional authority that bolstered belief in the presence of bitumen, sulfur, and similar substances beneath the earth's surface. The second book of Pliny's *Natural History* became a source for lecture material in Lutheran universities in the years after Melanchthon's reforms. Commentaries on this work addressed both Pliny's mentions of naphtha and naphtha's role in subterranean fires. Milich, in his 1543 commentary, explained that "naphtha is a Persian word that means a very strong flow of bitumen—not unlike the *amurca* of oil—which is mixed with sulfur." After explaining how nothing burns more than this substance, he concluded that subterranean fires occur where the "earth is fat with sulfur, niter, alum, or another combustible stuff, since when mixed with unctuous bitumen it can quickly inflame and burn for a long time."[35]

RENAISSANCE ARISTOTELIANS

While bitumen and sulfur increasingly played important roles in understandings of natural fires and the actions of the subterranean exhalations among writers who were working outside of Italian universities, the influence was slow to move over to commentaries on Aristotle's *Meteorology*. By the middle of the sixteenth century, a number of commentators were reading Aristotle in the original Greek and were more concerned with interpreting the work historically than merging their reading of Aristotle with new sources and observations. As a result, few commentators linked mineralogical views to explanations of aerial phenomena, as had Albertus Magnus. For example, Agostino Nifo discussed the causes of thunder and lightning at length, expanding on the nature of the hot and dry exhalation without referring at any time to bitumen, sulfur, or any other mineral or product of chymistry.

Nifo's explanation of thunder and lightning relied on his distinguishing the qualities of the exhalations that course through the middle region of the sky. He hypothesized that the hot and dry exhalations in this region are of two kinds. One is light and rare; it moves toward the upper part of the sublunary region where it either vanishes or bursts into flames. The other exhalation is thicker and heavier and becomes enclosed in clouds, which are composed of vapor. The coldness of the clouds causes the enclosed exhalation to become denser, through the process of antiperistasis, which causes it to strike the sides of the cloud, splitting the cloud and thereby producing a sound.[36] While in this explanation Nifo used density and rarity as the key *explanans,* in the description of lightning, he elaborated the effects of these qualities. In negative terms, he rejected both the view that lightning is the flashing of the stars (*fulgor astrorum*) and the view that lightning is composed of the hot and dry exhalation mixed with the unctuous—seemingly rejecting the possibility that lightning is bituminous. Rather, for Nifo, the matter of lightning is the hot and dry exhalation. When enclosed in clouds the exhalation thickens and intensifies in its heat either through antiperistasis or because the most subtle parts of the exhalation are raised to the highest area of the sublunary region by the power of the sun and stars, leaving behind only the thicker parts. These thickened parts, having been agitated by the neighboring vigorous heat, then begin to burn, thereby causing lightning. The strength of the lightning depends on the relative density of the exhalation and the cloud: the denser the exhalation, the stronger the lightning. A thin exhalation causes white light; a denser one produces yellow light; and, if mixed with something dense and wet, the burning exhalation emits red-colored flames.[37] Thus Nifo used only two qualities, density and rarity, to explain a range of phenomena, without relying on bitumen or unctuous moisture.

Similarly Francesco Vimercati, a generation older than Nifo, contended that the density and rarity of the dry and hot exhalation, when enclosed within the concave shape of the cloud, were responsible for variations in thunder.[38] And while he noted the similarity between thunder and the sound made by cannons when a "shell is discharged by the force of an ignited exhalation," he did not pursue the comparison so far as to find an analogy between the ammunition for the cannon and the material constituents of the dry exhalation, believing that there were fundamental differences between cannon fire and the rupturing of clouds.[39]

Some Aristotelian natural philosophers, however, recognized the importance of sulfur and bitumen in meteorology even if they did not identify these minerals with the dry exhalation. The preeminent example is Simone Porzio, a Neapolitan professor at Pisa during the 1540s and '50s. In a letter to Pietro Toledo, the viceroy of Naples, Porzio set out to give the natural causes of the extraordinary event of 1538, when the *campi flegrei* near Pozzuoli erupted into flames and gave rise to a new mountain, Monte Novo. Porzio began his explanations by appealing to commonly known features of this region: its vicinity to the sea; its abundance of hot water and sulfuric mud; its mountains filled with caverns that made this region prone to earthquakes, including the one that had occurred just two years before.[40] Porzio admitted that some things that occur only rarely "lack certain and definite causes," but others do have definite causes, such as strange lights, fiery exhalations, and earthquakes, regardless of their infrequency.[41] His explanation of this conflagration relied on the commonly perceived character of the area around Naples with the accepted actions of the dry exhalations and bitumen.

Following Aristotle, Porzio recognized vapor and the dry exhalation to be the material substrates of meteorological phenomena. The hot and dry exhalation, which he referred to as smoke, is generated in subterranean hollows and caverns. Some of it escapes through the terrestrial pores and generates the winds. The rest of it is trapped within underground veins.[42] In the case of the *campi flegrei*, its numerous pores and caverns cause large quantities of the dry exhalation to collect beneath its surface. The quick motions of exhalations ignite the bituminous material present in this region. After having been "burnt by fire, closed within the earth's hollows, [the exhalations] eject it [the bituminous matter], having been propelled by a great force."[43] The exhalations are distinct from the bitumen and act as the efficient cause, moving the earth by intensifying the fire, which Porzio contended was already burning, as is evident from the hot waters that gush out from this region.[44] While Porzio made his case through appeals to common knowledge, the analysis based on mineralogy is vague and the presence of sulfur and bitumen is stated as fact, but further analysis of why these substances are apt to burn or how they differ is wanting.[45]

In contrast to mid-sixteenth-century Aristotelians, those who at least partly defined themselves by opposing what they believed was the status quo were more likely to employ chymical explanations. Girolamo Cardano in his *De subtilitate*, for example, identified the matter of earthquakes as "what is suitable to burn," namely, "sulfur, sal niter, or hal niter, and bitumen." These were both the material substrate and the efficient cause of earthquakes, for "when they burn, and do not find a way out, just as in mines and bombs, they move the earth and shake." He even ranked the substances by their power to move the earth, with sal niter being the strongest and sulfur being the weakest.[46]

GIOVAN BATTISTA DELLA PORTA

The idea that the material substrates of subterranean and aerial fire were related was available through commonly read works, such as Pliny's *Natural History*, the Aristotelian *Problemata*, and Albertus Magnus's *Opera*, throughout the sixteenth century. Nevertheless, while a number of authors were expanding on the role of bitumen, sulfur, alum, and sal niter in mineralogical works, the wholesale importation of these fires to the upper regions of the atmosphere would have to await Giovan Battista della Porta's *De aeris trasmutationibus*. This work, first printed in 1610, was based on vast erudition. Its sources included Pliny, Simone Porzio, the *Problemata*, and large selections of Greek and Latin literature.

Della Porta's work on meteorology differs from traditional Renaissance ones in several respects. The kinds of differences should not be surprising considering his body of work. He was an avid proponent of experimentation, a member of the *Accademia dei Lincei*, and an author of numerous works on diverse topics, among them natural magic, the telescope, physiognomy, the art of memory, and distillation. In line with della Porta's commitment to practical endeavors, *De aeris transmutationibus* mixes considerations of causation with discussions of prognostication and predictive signs, a mix that is present to a much more limited degree in Aristotelian commentaries. His concern with utility and his application of findings taken from his investigations into distillation and refraction conform to his image as a participant in the creation of an alternative to

Aristotle's meteorology. His propensity, however, to quote a plethora of Greek and Roman poets, historians, and naturalists as part of his presentation makes him seem less a harbinger of something new than a devotee to a crepuscular version of Renaissance humanism.

His familiarity with ancient authors likely added to his willingness to connect the subterranean with the aerial. This willingness was explicit and thorough: "An earthquake is nothing other than subterranean thunder, and thunder is an earthquake in the sky." [47] Earthquakes and thunder possess the same underlying cause, that of burning sulfur and bitumen. He attributed earthquakes to underground fires that create excessive amounts of vapor (*spiritus*) that eventually shake the earth, spitting out the enclosed air aboveground. His explanation was by no means novel. He wrote, "Subterranean fire, acting in bitumen and sulfur and other inflammable liquors, sucking out a smoky and fatty exhalation, which is carried through underground caves and there, closed up, is agitated, just as gunpowder that is lit in cannons, or as vapor closed up in siege mines or in many other vessels that upset and knock down fortresses, camps, and cities." [48] Thus his explanation emphasized the unctuous and drew on analogies from the artificial. He based his views on evidence taken from Pliny as well as from unnamed witnesses of locales prone to earthquakes, such as Pozzuoli, Naples, and Abano, where underground fires emitted detectable fumes.

While his explanations for earthquakes corresponded to some of his contemporaries' explanations, his approach to fires in the sky was novel in that he defined one part of the sublunary realm, the area at the border between the two upper parts of the sky, as the bituminous region. The phenomena of this region are fiery and include many categories, either taken from Greek terminology or from Pliny's encyclopedia. The phenomena are grouped according to their composition, their shape, whether they are real or apparent, and the extent to which they move. They include lightning and thunder; meteors and meteorites with shapes that resemble streaks, beams, trees, disks, and stalks; colored lights in the sky called *chasms*, *pits*, and *bloody skies*; falling stars and dancing goats that hop across the sky; Saint Elmo's fire; and whirlwinds such as hurricanes, cyclones, and tornadoes. [49]

Della Porta provided an alternative material cause for Aristotle's dual exhalations, which he believed better explained the variety of meteorological phenomena. He contended that plants and the earth contained respirable humors and juices. When the sun heats pure water, it becomes air, which, when it condenses, produces only rain. If other "liquors" are heated, they create different kinds of breaths and fumes, including "fatty sulfuric" and "bituminous and

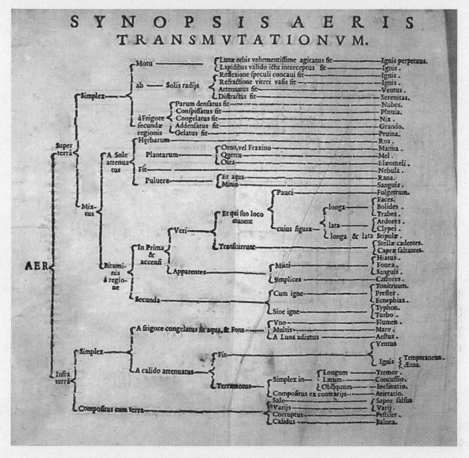

From Giovan Battista della Porta, *De aeris transmutationibus* (Rome: Zanetti, 1610).
Courtesy of the History of Science Collections at the University of Oklahoma.

inflammable" exhalations. These exhalations "retain all the powers and qualities" of their source.[50] Using this framework, della Porta was able to base explanations of what is not easily observed on the properties of more familiar things.
Thus he explained the bituminous region as being easily ignited because "the
sun absorbs rarefied fat from resinous and oily trees, bituminous locales, and
other things, and carries it up into the sky toward the torrid and burning ex-
panse of the heavens. There, in that spot, they catch on fire and transform into
various and diverse guises."[51] Because della Porta considered these phenomena
to be identical in nature to terrestrial bodies, he was free to employ analogies to
artificial processes and chymical reactions. For example, he concluded that the
colors of so-called chasms and bloody skies are similar to the hues of artificial

"amusing flames" that brighten holiday nights. In both pyrotechnics and the upper region of the sky, purple comes from burning sulfur, red from powdered charcoal, blue from sal ammoniac, and green from rusting copper.[52]

Similarly, della Porta's knowledge of gunpowder allowed him to dismiss as an old wives' tale the theory that thunder is caused by vapors escaping from clouds. While excusing the ancients, he contended that moderns, who have seen firearms, should know that thunder does not occur necessarily within clouds; cannons produce similar explosions yet are not shadowlike (*umbra-tilem*). Della Porta gave two alternative explanations for thunder. In the first scenario, fumes from manure, the sweat of animals, and materials such as sulfur, bitumen, and other fatty minerals rise up by the power of the sun's rays into the area that borders the aether, where it ignites because of the region's natural heat. In the second scenario, della Porta relied on his knowledge of refraction, contending that, because of the convexity of the outer edge of this part of the atmosphere, the sun's rays focus on the exhalation and ignite it, in a manner similar to using a lens to start fires artificially, something which he explained in his *De refractione*.[53] In either case, the burning exhalation then breaks apart the clouds, which are blocking it from the surrounding looser air. Because of the cloud's large size, its destruction necessarily produces a loud sound, namely thunder. The exhalation, however, need not be closed up inside the cloud. Della Porta explained that a parallel could be found in recent accidents in which gunpowder—composed of sulfur, sal niter, and charcoal—exploded, making loud sounds, when exposed to air and sunlight. These events demonstrated to della Porta that the noise, of both thunder and cannon fire, corresponds to the explosion of the fuel, not the narrowness of the container.[54] Thus evidence of the inadequacy of previous explanations of thunder is found in new well-known experiences with artificial combinations of chymicals. But behind della Porta's explanations is the conviction of the existence of unified causation for similar appearances, whether those appearances are subterranean, artificial, or high in the sky.

LATE ARISTOTELIAN CHYMISTRY

Seventeenth-century Aristotelian natural philosophers were far more likely to consider chymistry in positive terms and to integrate chymical explanations with traditional ones than their immediate Renaissance predecessors were. This intermixing of traditions, however, was by no means new. Medieval authors of chymical treatises, such as Pseudo-Geber, relied on Aristotelian frameworks, and medieval natural philosophers, such as Albertus Magnus, utilized the empirical findings of chymists in discussions of minerals. The growing importance of chymistry and its operational methods during the sixteenth century broke the relative isolation found in sixteenth-century Italian universities, among other places, where chymistry was discussed largely in the abstract. By the beginning of the seventeenth century, Aristotelian commentators considered the possibility that the fourth book of the *Meteorology* was foundational for chymistry,[55] and chymical practitioners were tapping into corpuscular explanations that ultimately had a basis in that same book.[56] At the turn of the seventeenth century, philosophers and self-defined chymists alike saw the fields of meteorology and chymistry moving in concert. The Jesuit Coimbran commentary on the *Meteorology*, for example, despite its skepticism toward the chrysopoetic claims of chymists, contended that lightning has a sulfuric odor because the hot and dry exhalation comes from sulfuric parts of the earth.[57] Andreas Libavius, the Saxon author of many chymical writings, noted approvingly that philosophers attribute fires in the sky to sulfuric, bituminous, and nitric exhalations that are ignited either by the process of antiperistasis or by friction.[58]

One of the most prominent figures in this shift was Daniel Sennert, a professor of medicine at Wittenberg from 1602 to 1637. His laboratory testing, which relied on a method called the *reductio in pristinum statum*, provided some of the best observational evidence for the theory that corpuscles, each of which retained their substantial forms in his view, were the underlying matter of all mixtures. According to Sennert mixtures could be reduced artificially to their initial components and then recomposed; their creation and destruction was the result of *sunkrisis* and *diakrisis*, the very Greek terms for *combination* and *separation* that Aristotle used in his explanation of cloud formation.[59] Sennert's reading of the *Meteorology* and his commitment to chymistry laid the foundations for his chymical explanations of meteorological phenomena in his *Epitome naturalis scientiae*, one of the most frequently printed textbooks of natural philosophy during the first half of the seventeenth century. Sennert's employment of chymical causation differed from della Porta's. According to Sennert, there is no distinct bituminous region of the sky. Sennert is not self-consciously rejecting Aristotle. There are similarities, as both della Porta and Sennert found that the knowledge of physical transformations that were easily observed could be transferred to explain changes that occur in the upper atmosphere.

Sennert offered several innovations to the understanding of the exhalations. He contended that there were two kinds of hot and dry exhalations. One type is *ek phlogosis* or *phlogiste* (Greek terms that refer to inflammability). This terrestrial exhalation, it appears that it retains the substantial form of earth because "the exhalations of salt of tartar, sulfur, and an infinite number of metals preserve their former species intact" in chymical tests.[71]

Since the methods of chymistry gave evidence for the persistence of substantial forms in vapor and the terrestrial exhalation, Froidmont could apply the properties of chymicals in his explanations of meteorological phenomena. He maintained that thunder makes a crackling noise when a burning spirit within a cloud is rarefied suddenly and breaks up the cloud. This kind of thunder is similar to the explosion of powder in gunfire. It is also akin to the sound of fire in a coal pit; it is produced by a "charcoal and bituminous exhalation."[72] Froidmont pursued this analogy in a detailed comparison of gunfire and lightning. He discussed a variety of recipes for gunpowder and the elevations that cannon fire can achieve. Basing his argument, in part, on the work of Libavius, Froidmont contended that the matter of lightning was similar to that of gunpowder, putting forth a theory identical to Sennert's that lightning was nitrous and sulfuric. Evidence for this is found in the supposed sulfuric odors of lightning and the quickness with which niter burns. Other chymicals gave potential explanations for the zigzagging motion of lightning.[73] The presence of burning gold powder, or *calx Martis,* explains lightning's downward direction, and niter explains its horizontal direction. Froidmont also thought he could explain away apparent differences between lightning and gunpowder. Ancient sources observed that lightning penetrated only five feet into the earth, while gunfire went much deeper. According to Froidmont this difference resulted from the great distance between the clouds and the earth, which diminished lightning's ability to penetrate.[74]

Froidmont's application of chymical methods and concepts and his hostility to some aspects of the chymical tradition resonated in the works of other Aristotelians, including those with a conservative bent. Bartolomeo Mastri was a Franciscan who, several decades before becoming the vicar general of the order, lectured on philosophy at Padua with his fellow Franciscan, Bonaventura Belluti. In between the years 1637 and 1647, the pair published seven volumes of commentaries on logic, natural philosophy, and metaphysics. They explicitly aimed at defending Scotist thought.[75] The two Franciscans realized that defending a comprehensive Scotist meteorology posed difficulties because the questions on the *Meteorology* that are contained in the Wadding edition of Scotus are spurious while Scotus's extant writings focus more on metaphysics and logic than on natural philosophy.[76] Thus their examinations of the sublunary world necessarily depended on extrapolations that went beyond a strict devotion to Scotus, though they remained true to that tradition as much as possible. As a result, a hesitancy to accept the new and an adherence to official church positions emerge at times in their writings, as is evident not only in their rejection of the recently condemned views of Galileo regarding the motion of the earth but also in their support of the view that comets are sublunary, a view that they presented in *De celo et meteoris* (1640).[77]

They found, however, no danger in utilizing some of the chymical techniques of investigating meteorological phenomena, even if they rejected the possibility that chymists could artificially make gold and instead maintained that substantial forms could only come from the heavens.[78] Mastri and Belluti proposed that a Paracelsian theory for the production of metals—in which sulfur acts as the father implanting a seed in mercury, the maternal matter—is at least somewhat probable. While experience gave some support to the Paracelsian view, in the end they sided with a position they attributed to Aristotle and Albertus Magnus that held that metals are mixtures of water and earth.[79] They asserted that the terrestrial exhalation burns because of its inherent oiliness and that the material cause of subterranean fires and volcanoes is sulfuric and bituminous—beliefs that at the time conformed to accepted opinions. The latter position was further supported by observations of the 1631 eruption of Vesuvius.[80] Their knowledge of chymical procedures did not stop with a sympathetic recapitulation of basic Paracelsian theory and the endorsement of the role of bitumen and sulfur in the creation of subterranean heat. They believed, just as Froidmont had, that experimentation could help establish what happens during the creation of the exhalations.

Willing to admit the probabilistic nature of meteorology, Mastri and Belluti thought that artificial interventions into nature could help decide difficult questions. One such question was whether "vapor differs in species from water." They drew a parallel between the vapor that arises from rivers and seas and the vapors that are extracted from herbs through distillation. This comparison led them to conclude that it was "more probable" that vapor possessed the same substantial form as water. When herbs are heated they form two different liquid substances: one that is a kind of juice or humor, the other that is aqueous. Once separated these two liquids behave differently. The juice, despite retaining the properties of the herb, will not revert back to its original state. On the contrary, the aqueous part "right away reduces itself to the pristine form of water, which we do not experience in the juices of herbs." The same is true for liquid mercury. The two Franciscans concluded that "because just as quickly as the vapor is extracted, it is as easily converted back into water, these mutations cannot be

CHAPTER FIVE

Niccolò Cabeo's Meteorology as the Basis for a New Aristotelianism

By the middle of seventeenth century, a symbiotic relationship had developed between chymistry and Aristotelian meteorology. As shown in the previous chapter, natural philosophers applied knowledge of substances such as bitumen and sulfur to their explanations of aerial phenomena. Niccolò Cabeo pushed the limits of this relationship, as he contended that the *Meteorology* should be the textual basis for investigations into the natural world, investigations that were modeled on the practices of chymists. Cabeo sought to reformulate Aristotelian natural philosophy by emphasizing laboratory experiments and corpuscular matter theory while simultaneously minimizing the role of metaphysics. This reformulation found its basis in the study of meteorology.

Cabeo's emphasis on laboratory experiments and corpuscular matter theory in his work on meteorology parallels some of the major developments of chymistry at this time. Among the most relevant of these developments was the work of Daniel Sennert. Relying on a tradition of experimentation that went back to Pseudo-Geber, Sennert thought laboratory investigations could illuminate matter theory by showing that macroscopic objects were composed of small corpuscles. The chymical tradition, in which Sennert took part, relied on Aristotle's *Meteorology*; the tradition found authority for corpuscular views in Aristotle's description of the formation and dissolution of clouds via *sunkrisis* and *diakrisis* (1.9.346b21–35) and in the discussions of matter theory that equated artificial processes to natural ones (4.3.381b1–8) and used pores to explain the passive qualities of matter (4.8.385a28–32; 4.9.386b2–11; 4.9.387a2–4; 4.9.387a19–21).[1] Cabeo was aware of this chymical tradition. Although he did not cite

Sennert, he nevertheless relied on Pseudo-Geber's corpuscular matter theory and described chymical experiments and tests that he presumably conducted.

The emergence and acceptance of corpuscular and atomistic theories of matter are among the more important changes of seventeenth-century natural philosophy. Numerous proponents of new natural philosophies who critiqued Aristotle were also proponents of corpuscular explanations. Galileo Galilei and René Descartes were among those whose works preceded Cabeo's and promoted corpuscular matter theory. It is unclear whether Cabeo was familiar with Descartes' works, even though there are similarities in their larger purposes. Cabeo had read some of Galileo's works and held a negative view of portions of the *Dialogue Concerning the Two Chief World Systems*.[2] Cabeo does not appear to have been influenced by the corpuscular matter theory put forth in the *Assayer*. Nevertheless, Cabeo's natural philosophy was part of the tradition that mixed Aristotelian thought with chymistry and thus played a role in the emergence of corpuscular matter theory.

Cabeo (1585–1650) was a Jesuit who taught and studied in Parma, Mantua, Bologna, and Ferrara among other cities in northern Italy, working both in colleges as well as in courts.[3] He was successful in both arenas and went from being known as a child prodigy to eventually gaining a reputation as a *"grand' huomo,* famous in print." Cabeo's learning gained him entry into Ferdinando Gonzaga's ducal court, where he was on familiar terms with the duke himself.[4] Cabeo's fame derived from his two published works. The *Philosophia magnetica* (1629) is best known for its anti-Copernican stance, its explanation of magnetic and electrical phenomena as caused by effluvia, and its attacks on William Gilbert's contentions about the magnetism of the earth.[5] His second printed piece, the *Commentaria in libros Meteorologicorum* (1646, reprinted 1686), discusses a wide variety of subjects as well as Aristotle's text in its lengthy four volumes. That Cabeo's only commentary on Aristotle, and his largest work on natural philosophy, was dedicated to meteorology and the *Meteorology* is significant.

Cabeo's advocacy of corpuscular matter theory and chymical interventions into nature underpinned a reform of Aristotelian natural philosophy that was based on his interpretation of the *Meteorology*. This reformulation took into account contemporary theories and practices but retained reading Aristotle as foundational. As a result, his natural philosophy appropriated recent discoveries yet could remain palatable to conservatives. First, like many seventeenth-century philosophers, he decried what he understood to be the dependence on and faith in Aristotle's authority in scholastic circles. Second, he maintained that *physica* should be free of metaphysical and mathematical explanations. In

Cabeo's eyes, because Aristotle was too occupied with metaphysics, dependence on his writings caused other Peripatetics to ignore physical objects or to analyze nature using only metaphysical concepts. In this respect he opposed the style of philosophy and discussions of his Jesuit predecessors, such as Suárez, Toletus, Fonseca, and the Coimbrans, all of whom stressed metaphysics.[6]

COMMENTARY AND *PHYSICA*

Cabeo's attacks on Peripatetics were made complex by the fact that he insisted, despite some reservations, that an accurate understanding of Aristotle's text, in this case the *Meteorology*, was foundational. Since philosophical conclusions, according to Cabeo, should be independent "from authority" and based on "evident reason,"[7] true philosophy must go beyond the acts of "reading (*lectio*), commentary (*commentatio*), or speculation (*speculatio*)."[8] Indeed, he believed other commentators had been led astray by trying to solve Aristotle's obscurities without personally experiencing nature and thus had adopted a religion (*fides*) of Aristotelianism rather than a science (*scientia*) based on observation.[9] He was particularly wary of bibliolaters who thought that establishing the meaning of the text is the final goal. These adherents to Aristotle thought that "Aristotle always tells the truth" and that "it is a kind of blasphemy to affirm that Aristotle did not tell the truth."[10] Cabeo's description of Peripatetics is perhaps hyperbolic, having more in common with Galileo's fictional Simplicio than with prominent Peripatetics of the day. The previous chapter demonstrated that Aristotelians integrated findings taken from nontextual or non-Peripatetic sources in their formulations of meteorological theory. Even Cesare Cremonini, the Paduan philosopher who gained notoriety for what he claimed to be literal interpretations of Aristotle and for ensuing run-ins with the Jesuit order, did not maintain that all of Aristotle's precepts were true. In fact, in the prefaces to his commentaries on the *De caelo* and *De anima*, Cremonini wrote provisos that these interpretations were of Aristotle and not necessarily what Cremonini thought to be true, although he possibly included these warnings to escape ecclesiastical prosecution.[11]

Instead of blindly assenting to Aristotle's claims, Cabeo engaged in a paradoxical exegesis, whereby Aristotle's words themselves were evidence that he should be doubted and that his writings could not be trusted absolutely. The text of the *Meteorology* is crucial to this point because, as discussed in the first chapter, Aristotle wrote in several places that his solutions were provisional and not demonstrative. Cabeo realized this and wrote, "He asserts that in many topics he doubts; that is, he answers the issue in question by asserting that nothing is for certain. Therefore not everything is demonstrated, or rather, it is not as though everything is put forth as determinedly true."[12] Aristotle's conviction that many of his positions were not demonstratively true placed the burden on his readers to evaluate and potentially improve his doctrines. Setting out from Aristotle's self-realization that some of his conclusions were at best probable, Cabeo deviated from traditional interpretations to a startling degree. He rejected, for example, the four elements and replaced them with Paracelsus's sulfur, mercury, and salt. Yet he did not see this replacement as necessarily contradicting Aristotle, whom it is possible to interpret in a wide variety of ways, depending on which texts and passages are emphasized.

Despite his reservations about other Peripatetics who dwelled on philological and linguistic issues, Cabeo stressed the importance of accurately interpreting Aristotle's text: "It is primary to know what Aristotle said on each issue, to understand thoroughly his words and the meaning of those words, and to conceive with the entire mind the arguments, which he uses to prove particular issues."[13] Thus a commentator should strive to clarify the text as it stands, not as it would be with additional words and phrases, which Cabeo maintained was unfortunately the custom.[14] The understanding of the precise meaning of the text is "primary" but not the destination. As a result, Cabeo divided his work into *lectiones* and then *quaestiones*. The former were intended to reveal accurately Aristotle's intent, while the latter aimed to go beyond what Aristotle believed and search out the truth unfettered by the demands of ancient authority and bolstered by knowledge derived from the senses. Aristotle's doctrines are not "dogma brought down from the sky," but if "one wants to advance toward knowledge, one must examine the standard of truth, discuss his uncloaked propositions very accurately, expound the force of his reasoning, and expose the causes brought forth from Aristotle himself."[15]

Cabeo's evaluation of Aristotle was most negative with respect to the integration of metaphysics and natural philosophy. In his eyes, many of Aristotle's views could not be accepted because he was "more accustomed to metaphysical speculation than physical observation."[16] In the class of metaphysical speculation, he included abstractions and the unnecessary deployment of scholastic logic that depended on reducing things (*res*) into universal categories, differences, and divisions, all three of which have no physical reality. He rejected metaphysical entities as chimerical because they are neither material nor physical nor sensible.[17] These critiques of metaphysics were known among later more vehement attackers of Aristotle. Robert Boyle, who apparently read the *Commentaria*, called Cabeo an "inquisitive Peripatetic" and "one of the most judicious" commentators, noting that faithful adherents to Aristotle used metaphysics as an interpretative crutch by claiming that Aristotle shifts between "*actu & potentia* . . . to shift off those Difficulties he could not clearly Explicate."[18] As a result, dishonest sophistry and blind faith in Aristotle were to blame for the frequent metaphysical interpretations of natural philosophy.

According to Cabeo, reasoning (*ratio*) and especially experience of sensible causes were essential for making progress in natural philosophy and for freeing oneself from the dogmatic faith of Aristotle.[19] This concern is common to his earlier work. In the preface to the *Philosophia magnetica*, he wrote that he did not think "that any person can sustain himself worthy of philosophy, even if he should muster up some metaphysical subtleties, unless he should have accumulated in his mind the physical and sensible causes of these things, which are produced daily by nature."[20] He intended to correct both commentators of Aristotle who were consumed only by reading and Aristotle himself who was consumed by metaphysics.

Cabeo separated *physica*, as he called the study of nature, from the other two subjects he considered part of speculative science: metaphysics and mathematics.[21] For him *physica* concerns the sensible properties of physical bodies and their effects, that is, "all effects of those things that can be perceived" and "the sensible causes of all effects, which can be perceived by external sensation," while "those [causes] that cannot be perceived do not pertain to *physica*."[22] Equating the real with physical and material objects, he maintained that metaphysical categories have no bearing on our understanding of the natural world. Because the goal of *physica* is to explain sensible causes and effects, Cabeo largely appealed to observation, experience, and contrived experiments. The experiential emphasis of the *Commentaria* was not lost on his contemporaries. Marin Mersenne stated in a letter written to Athanasius Kircher shortly after the *Commentaria*'s printing, "What [Cabeo] wrote on the *Meteors* is certainly beautiful, for he does not follow the commonplaces of philosophers, but he judges what is best according to experiences."[23]

His rejection of the application of metaphysics to nature and his endorsement of experience and experimentation led Cabeo to be particularly concerned about the prominent position of substantial forms in Aristotelian natural philosophy. He believed the common understanding of form as essence was mistaken and untenable for discussions of nature, since forms are insensible and nonphysical. Rather if forms were to be admitted into discussions of nature they would have to be real, physical, material entities. He dismissed

Aristotle's understanding of substantial form because it "is a metaphysical essence and formula according to Aristotle; it is not a physical entity,"[24] and therefore it has no role in natural philosophy. The twin concepts of form and privation, as traditionally understood—"one of which is nothing, the other which is metaphysical"—were also rejected.[25]

As an alternative, Cabeo posited substantial forms to be spirits and vapors, endowed with various powers and virtues. Redefining form as active and physical, he wrote, "Form is truly physical, it is a spirit, vapid and subtle; for it is that which gives determined being to each thing. For, there is such a thing that is animated by this kind of spirit. From this [spirit], there is an active force, so great and of such kind; and just as the diversity of the sublunary objects comes from these spirits, which are implanted in them, the diversity of faculties, properties, and virtues comes from these. This is true act, this is true form, not metaphysical, a theory conceived in the mind, but a physical principle of faculties."[26] What Aristotle called form, and what some consider metaphysical, is in fact a specific kind of body that unifies the substance. It is a spirit, a vapor that consists of small parts. Like the Stoics, Cabeo's forms are active and entirely material forces that order the world and its contents.[27]

While Cabeo's particular views about substantial forms might have been innovative among Aristotelians, the questioning of their role in meteorology was common to at least some of his contemporaries. John Poinsot (also known as John of Saint Thomas) in his "Tractatus de meteoris," which was part of the 1634 *Cursus philosophicus thomisticus*, began with the proclamation that he would not apply final and formal causes, only material and efficient ones.[28] The efficient cause is divided into two: *per se,* which is the power (*virtus*) of the sun, stars, and celestial bodies that comes in the form of heat; and *per accidens,* which is antiperistasis. Poinsot utilized corpuscular motifs to explain how heat acts as an efficient cause. According to Poinsot, vapor is composed of subtle aqueous parts. Heat causes evaporation by lifting these subtle parts to the higher regions, where they in turn fall, causing precipitation. The hot and dry exhalations, however, are affected by the heat and *virtus* of the sun and stars, which make them rise and eventually flame up, causing winds, thunder, lightning, and comets. The power of heat acts by thinning out (*subtilizando*) both vapors and exhalations and by separating (*segregando*) the more subtle parts from the thick ones. After these subtle parts have reached higher levels above the surface of the earth, they either burst into flames or precipitate depending on whether they are smoky or watery.[29] Thus in Poinsot's textbook meteorological phenomena are caused by the separation and motion of small particles without

The innovations in Cabeo's matter theory did not end in his rejection of the traditional understanding of substantial form; he also rejected the four traditional elements and integrated the Paracelsian *prima tria* with the more Aristotelian earth, water, and spirit. He contended that his adoption of the three elements was not novel or a refinement of Aristotelian theory but that it came from Aristotle himself. Cabeo wrote, "This is true Peripatetic doctrine, here expressly handed over by Aristotle, so that you do not think that I have forged a new philosophy; [rather, I] repeat the very thing from Aristotle, which Aristotle took from the ancients." [47] The evidence in this case is textual, as Cabeo took a unique interpretative position. The supporting passage, which comes from the fourth book of the *Meteorology*, reads, "Therefore, the homeomerous bodies both in plants and animals consist of water and earth. And what are metals, such as gold, and silver, and whatever else of this sort [consist] of these [water and earth] and an exhalation, either of which, is enclosed [within the earth] as is stated in other places." [48] Equating exhalations with spirits, and taking the phrase "homeomerous bodies" to mean "all homeomerous bodies," he concluded that all sublunary substances are composed out of earth (fixed parts), water (the wet medium), and the double exhalations (spirits), and that these are the "true elements." Exiling air and fire from the pantheon of primary constituents, Cabeo wrote, "I think that Aristotle, whatever other interpreters say, spoke universally about all things, not just about some, and he concluded that all homeomerous bodies are made of four bodies, earth, water, and the double exhalation." [49] For Cabeo, the material substrata for Aristotelian meteorological theory become the elements for the entire real, physical sublunary world. Vapors and exhalations are the spirits that define the diverse number of sublunary bodies through active powers. [50]

While textual explication underpinned his reframing of the Aristotelian elements, more often Cabeo appealed to the sensible world to support his corpuscular matter theory. Albeit, knowledge of spirits is gained from experience, not experimentation. Vegetable spirits unite the material of plants and give it life, just as animal spirits, "as experience establishes," give life to animals. [51] Upon death these spirits break up and separate themselves from the body, although Cabeo cautioned that "it is not possible to say that the [intact] soul of a horse or a dog flies away; rather, the spirits having been broken-up move away and disperse." [52] The presumed existence of spontaneous generation further supports these claims, according to Cabeo. He argued that animals that are produced without seed are created in dung, which contains animal spirits that had been separated from animals. For example, worms are generated out of rotting meat, which is "filled with animal spirits" that still adhere to the carcasses of organisms. [53]

Cabeo's appeal to quotidian experience in establishing the existence of vegetable and animal spirits is not indicative of the varied range of observations and experiences he utilized in his *Commentaria*. Experience for Aristotle and for most Aristotelians was a broad category that included commonplace or everyday experiences, which included observational facts that were thought to be accepted universally. Additionally the category of experience included other people's experiences or accounts of them in authoritative books; an experience need not be based on the author's own observation. Cabeo, however, discounted the role of everyday experience (*experientia quotidiana*) and chastised Aristotle for his dependence on it. Instead he claimed, "A philosopher will never become a scientific physicist (*physicus scientificus*) only by reading books of philosophy, unless he considers nature and takes up experiments (*experimenta sumat*)."[54] Thus he urged natural philosophers to rely on their own observations and actively experience and intervene in nature instead of relying on experiences described in texts.

Both Renaissance Aristotelians and seventeenth-century Jesuits have been credited with promoting empiricism in natural philosophy. Eckhard Kessler has argued that the movement away from metaphysics toward experiential knowledge characterizes a number of thinkers participating in or influenced by Renaissance Aristotelianism;[55] and Peter Dear has located the rise of mathematical experimentalism in the works of various seventeenth-century Jesuits. Cabeo's reliance on chymical testing and experience differs significantly from both of these groups. Dear contends that a group of Jesuits, inspired by a reading of *Posterior Analytics* 2.19, emphasized the role of sensation in forming universal concepts that were considered the basis of scientific knowledge.[56] Among some of these Jesuits, this interest in experience and experimentalism became intertwined with a reliance on mathematical proof to determine certainty. Again, Cabeo does not fit well with this group. First, while maintaining the importance of induction, he rejected the importance of using logic and forming universals in *physica*, thereby limiting the need to appeal to the *Posterior Analytics*. Second, Cabeo distrusted mixing the mathematical and sensible worlds.

Instead Cabeo's empiricism derived from the chymical tradition. Both medieval and early modern practitioners of alchemy had developed techniques of assaying and experimenting that were integrated with Aristotelian concepts.[57] It is in this tradition that Cabeo urged his scientific physicists to take up experiments so that by doing their own laboratorial research they could understand the works of nature. He used Baconian language to promote the use of experiments that would break away from bookish knowledge and directly confront the human soul. Unfortunately, the ramifications of this dualism are not discussed in the *Commentaria*, as he left those for discussions of "the books of the *De anima*," of which he authored none. What remains clear is that for Cabeo all substances from the lowest tissue up to but not including the human soul were generated from the position and combination of thin vapid parts, thick parts, in a watery medium.

DEFENSE OF FAITH

Cabeo's corpuscular and largely materialist understanding of nature ran counter to movements within the Jesuit order and some elements of the Catholic Church. It went beyond the scope of the metaphysically oriented philosophy as discussed in Suárez, Toletus, Fonseca, and the Coimbra commentaries. Moreover, the first half of the seventeenth century was marked by increasing concerns over atomism and related natural philosophies, particularly its implications for theological understandings of transubstantiation and the immortality of the soul. *Ordinationes*, issued by Jesuit councils in 1632 and 1651, forbade atomism and corpuscularism. As a result it became common for Jesuits to deny explicitly that substantial change was merely the congregation and separation of corpuscles.[69] Yet the open presentation of a physical system that explains physical change through the coming-together and dispersion of particles without recourse to the dichotomy of substance and accidents suggests that fears about corpuscular philosophies among Jesuits were not as far reaching as might be imagined. While censors noted that Cabeo's work espoused corpuscular views, they did not prevent publication.[70]

Despite his promotion of corpuscularism, Cabeo believed that his natural philosophy preserved the true faith. Experience not only corrects Aristotle but also affirms that Aristotelian natural philosophy is wrong on topics where it clearly deviates from tenets of Christianity. Cabeo attacked Aristotelians for their "faith" in Aristotle's words, a faith that was an impious and irreligious alternative to the Christian faith, he suggested. While considering the evidence for the creation of the world against Aristotle's position that the universe is eternal and was never created, Cabeo warned that "nevertheless there are men, who agree with that position [i.e., the world is eternal]; and they think that Aristotle demonstrates everything, and those men are enemies of the faith. Alas, that there be those in our time who follow Aristotle more than Moses."[71] Cabeo, however, did not maintain that biblical exegesis should replace Aristotle; rather, he wrote that truths from the Bible could be confirmed through his experiential practice of natural philosophy. He wrote, "Gladly I will see what the impious Peripatetics (*Atheistae peripatetici*) oppose, that which is physical, deducted from sensations, and based only on the light of nature."[72] Thus the light of nature is opposed to bookish faith in Aristotle.

Although it might seem obvious to present-day scholars that there are major discrepancies between some of Aristotle's positions and Catholic dogma, during the first decades of the seventeenth century, some university professors and Catholic theologians alike feared that the rejection of Aristotle might lead to a loss of authority or worse. Regents at the Sorbonne were particularly diligent in monitoring potential threats to Aristotelianism. In 1624 in Paris, Antoine de Villon, Jean Bitaud, and Étienne de Clave were censured for, among other positions, their advocacy of an alternative to the four Aristotelian elements.[73] The revitalization of Aristotelian thought during this period was due in part to the rise of the Jesuits. Cardinal Roberto Bellarmino believed that Aristotelianism should be promoted over Neoplatonism because the latter falsely appeared to be similar to the Catholic teachings, while it was more obvious where Peripatetic thought deviated from religious orthodoxy.[74] Accordingly, Jesuits became prominent leaders in commenting on Aristotle's writings.

The adoption of a Thomistic version of Aristotelianism by Jesuit and other Catholic authorities perhaps motivated Cabeo to justify his departure from traditional interpretations of Aristotle by branding the school of literal interpretation, which was associated more with secular professors of philosophy at universities in Padua and Bologna than with church officials, as impious. Who these unnamed impious Aristotelians were is not entirely clear, although some guesses can be made. The tradition of pursuing hypothetical arguments *secundum Aristotelem* that were against the doctrines of the faith is lengthy. The question of the eternity of the earth often played a significant part in these arguments. Thomas Aquinas contended that only through faith and not through rational argument could it be known that the world was created in time.[75] In the early part of the sixteenth century, Tiberio Russiliano, basing himself on Avicenna, argued that according to philosophical reasons the earth must be eternal, and therefore there had been an infinite number of cataclysms and recreations of life.[76] More recent, and undoubtedly more present in Cabeo's mind, was Cesare Cremonini.

Cremonini, a professor of philosophy at Padua from 1591 to 1631, objected to the establishment of a Jesuit college in the city on the grounds that it violated privileges that had been granted to the university centuries ago. The Jesuits brought him to the Inquisition twice, accusing him of teaching the mortality of the soul and the eternity of the world. Cremonini defended himself by saying that he was only explicating the true views of Aristotle, which he was required to do according to university statute. In the end, Cremonini was largely exonerated, and the Jesuits did not encroach too far on Paduan soil.[77] Despite the fact that in numerous works, both printed and in manuscript, he explicitly stated that the soul is immortal and the world finite in time, numerous scholars sympathetic to anticlericalism, since the time of Cabeo's death, have seen him as a thinly veiled cryptoatheist.[78] His Jesuit opponents most likely saw him as lacking in sincerity, and despite the fact that he had died in 1631, he perhaps lingered in the mind of Cabeo, who had been stationed in Padua at various times, leading him to posit the existence of numerous sympathizers.

Regardless of whether Cabeo's belief in the existence of these opponents was accurate, it remains to be seen how he thought that experiential and experimental corrections to Aristotle would result in the confirmation of Christian truths. His discussion of the marvelous and the seemingly miraculous emphasized that such phenomena should be understood by appeals to natural causes. Thus he denied that bloody rain could be explained by natural causes. While some wish to think that there was rain that was "true blood, generated from the physical world by natural causes, not with the higher will of God," Cabeo "could not admit that there was something, which stains plants and clothes with a purple-colored humor, which was true blood." Rather, the generation of blood requires a natural heat that is found only in animals. Thus bloody rain "[could] not be generated naturally or by the power of nature in clouds"; therefore, the accounts of bloody rain were false. He believed that "there was some red and purple humor, which, out of its similarity, the common people called it blood."[79] In this respect Cabeo is in step with the Jesuit order in general: the Coimbran commentary on the *Meteorology* took an identical position on bloody or milky rain.[80]

Similarly, the appearance of raining frogs is deceptive and false because it is impossible for there to be a sufficient amount of "fixed matter" in the clouds, which would be required to generate a mass of nebulous amphibians. Rather they are formed via spontaneous generation in the ground, and they emerge at the same time the rain falls to the ground.[81] Thus Cabeo, while not discounting the possibility of miracles, had no place for them in his discussion of the physical world. He distanced his study of *physica* from those concerned about prodigies and events that depend on divine intervention. The separation between the natural and the supernatural is most explicit in his discussion of flooding. Following the traditional distinction between particular floods (or local naturally caused floods) and universal floods, which could only be caused by God, Cabeo wrote that a "a philosopher does not dispute about universal [floods] because he does not know the causes of it in nature." Thus instead of discussing the issue *secundum philosophiam,* as Albertus Magnus and later Russiliano had, he contended that universal floods were supernatural, could not be caused within the natural order of things, and depended on God's free will; therefore, any difficulties must be treated by theologians, not natural philosophers.[82]

Cabeo's separation of physics from theology did not, however, lead him to think that observations of the physical world could play no role in confirming the truth of scripture. Specifically, he thought that physical evidence and historical observation showed that the universe, or more specifically, the earth, had a temporal beginning. His attack began by examining the ramifications of Aristotle's theory of the cyclical nature of not just life but also the features of the earth and human civilizations. In *Meteorology* 1.14, Aristotle contended that over time the sea changed places with the land, so what is now land will eventually become the sea and vice versa. Changes of the earth's topography are not unidirectional—the earth is not aging—but are cyclical, and by virtue of the infinite duration of the universe these cycles are likely to be infinite in number. Moreover, similar cycles of growth and destruction describe the history of human civilizations; ethnic groups have thrived, been destroyed in mass extinctions, and migrated. According to Aristotle, we are unaware of these past civilizations because their writings or other productions have been destroyed just as the land they lived on has been swallowed by the sea.[83] While this aspect of the eternity of the world and the eternity of its living species typically does not dominate discussions of the Aristotelian position, in a sense they are consequences of this doctrine. If it is conceded that humans have lived on earth for an infinite amount of time in more or less the same fashion that they do now, it follows that countless past civilizations, some of which we have no knowledge or evidence of, flourished and then eventually perished.

To Cabeo, Aristotle's positions were not only erroneous paths to impiety but also examples of Aristotle's willful disdain for history and the evidence. Furthermore, Aristotle's arguments for the mutability of the earth are at best probable and in no way demonstrative, contrary to what he considered the impious Peripatetic acolytes to have thought. According to Cabeo, they believed that whatever is found in Aristotle's texts is based on demonstrative proofs even if it contradicts religious truths. He recounted an apocryphal anecdote in which Aristotle, after obtaining a copy of the Old Testament and reading, "In the beginning God created heaven and earth," reacted by saying, "This author says

many things, but proves nothing." [84] Cabeo chastised Aristotle for not recognizing the Bible as a historical narrative that "does not require proof, but is a simple
exposition of facts." As a result, scripture becomes a storehouse for experiences
about the past, including experiences about terrestrial formation and possible
transformation. Thus Cabeo turns Aristotle's purported utterance on its head,
writing that someone should have responded to Aristotle by saying, "You say
many things, and you prove nothing." [85] The conviction that the seas and rivers
change locations has neither been proved nor is even probable, according to Cabeo. In this case Aristotle "does not treat the question more or less philosophi-
cally but poetically and according to an elegant manner for speaking." [86]

Historical experience reveals the implausibility of Aristotle's views, in Cabeo's eyes. He believed that while at times springs and small lakes emerge after
earthquakes, new seas never do. Rivers also appear to be stable: "In the last
4,000 years, there has always been the same Danube, Nile, Rhine, and Po." The
stability of land and sea is confirmed not just by scripture but by experiences
found in classical texts as well: "We have histories of nearly three thousand
years, both sacred and profane, which recount the seas that are small, when
compared to the Ocean: the Black, the Ionian, the Adriatic, the Caspian, [and]
the Red Seas. These seas, with the passing of so many centuries, have neither
changed, nor dried out, nor perished, nor grown, but have as a rule persisted, as
they were then." [87] Therefore, Cabeo believed that the evidence points to the
relative stability of the earth, which in turn suggests if not the finitude of its
existence at least the incorrectness of Aristotle's views.

Cabeo continued his attack by addressing terrestrial features. He believed
that there is a natural principle for the wearing down of mountains, that naturally there is a force that levels the earth's surface. Yet there is no evident contrary natural force that creates hills and mountains. Earth differs from water
by its lack of such restorative principles: there is a principle of desiccation and a
corresponding principle for restoring bodies of water through rain. But for
mountains, he concluded that they "diminish every day; and it is not possible to
imagine any way by which they are restored, therefore the earth was not from
eternity as it is now." [88] If the earth was eternal, all the mountains would have
been worn down, and the entire surface of the earth would have the same altitude. Therefore the earth must not be eternal or even extremely old. Character-
izing this argument as a "physical demonstration," he contrasted it with those
metaphysical arguments found in the eighth book of the *Physics*. Those are not
truly demonstrations, and the theories are not known per se or by the senses,
"as it is per se known that mountains diminish everyday but are never re-

stored."[89] Thus Cabeo attempted to use observations about the sensible world and physical evidence, rather than metaphysics or appeals to Church dogma, in his argument that the earth was created in time.

―∽ ∾―

In his efforts to remedy the metaphysical tendencies of his Aristotelian predecessors and peers, Cabeo argued that *physica* should be based only on physical principles. These principles largely came from the *Meteorology*. The spirits and vapors that traditionally explained aerial and subterranean change would then become the model for the entire natural world except for human psychology. Cabeo made the *Meteorology* a starting point in order to undermine metaphysical and speculative accounts of the natural world based on formal and final causes. Metaphysics having been removed from natural philosophy, knowledge of these physical principles could only derive from sensation. As a result, Cabeo prioritized the roles of observation, experience, and experimentation in natural philosophy, basing himself on the operational techniques and corpuscular explanations of chymistry. The elimination of metaphysical speculation led to a sharp division between many theological and philosophical considerations, since the latter specifically only concerned the natural and not the supernatural. As a result, Cabeo's *physica* mostly did not touch on religious issues. Nevertheless he believed that a consideration of meteorology and geology, aided by observations of the natural world, demonstrated the implausibility if not impossibility of an eternal universe. Cabeo turned experience against Aristotle, whom he believed was too metaphysical and not a careful enough observer of the natural world. This experience, in his opinion, not only helped establish a corpuscular matter theory against the traditional reading of Aristotle but also bolstered his critique against those who accepted Aristotle's allegedly blasphemous tenets based on an acceptance of literal interpretations of the text.

Cabeo's *Commentaria* shows the great flexibility of Aristotelians during the first half of the seventeenth century, particularly when they discussed meteorology. The field, with its emphasis on experience, material and efficient causation, corpuscularism, and chymistry, was ideal for Aristotelians, such as Cabeo, who wanted to divorce natural philosophy from its textual roots and reform it so it could be competitive with the novel natural philosophies that were becoming increasingly abundant, many of which relied on corpuscular explanations. In a sense, Cabeo's work is a culmination of the Aristotelian meteorological tradition. Using the *Meteorology*'s promotion of experiential evidence, its emphasis on material causation, and the provisional epistemological status of its subject, Cabeo dismissed what might be considered foundations of Aristotelian natural philosophy: the immateriality of substantial form and the importance of privation and form to explain change. Many of the novel natural philosophies contemporary with Cabeo's had similar goals and approaches. Increasingly, opponents of Aristotle found the metaphysical foundations unintelligible, unnecessary, and unconfirmed by experience. Descartes was perhaps the most prominent of these opponents whose ideas resonate with Cabeo's. In his work we can see how some of Cabeo's goals can be found in a contemporary who portrayed himself as rejecting entirely Aristotle's authority.

CHAPTER SIX

Causation and Method in Cartesian Meteorology

By the middle of the seventeenth century, a number of Aristotelians emphasized the provisional nature of meteorology and deemphasized formal causation. Moreover, some, notably Cabeo, not only used the behavior of corpuscles and small particles as key explanations but also found that meteorology could be used as a model for all of physics. It is against this background that parallels between René Descartes' *Les Météores* and the works of his colleagues and predecessors are best seen. More than seventy years ago, Etienne Gilson elaborated the similarities between Descartes' *Les Météores* and the Coimbrans' textbook that was based on Aristotle's *Meteorology*. The topics treated in Descartes' work correspond closely to those in the frequently taught Jesuit textbook. Both discussed the formation of clouds, precipitation, rainbows and other lights in the sky, the formation of minerals and salts, and the cause of winds.[1]

The similarities do not end with the structure and topics but extend to large portions of the content. In particular, there are broad correspondences between Descartes' views and those of Aristotelians toward the elimination of substantial forms from meteorology, the use of corpuscles to explain meteorological phenomena, the application of analogies to explain what is hard to observe, and the employment of a hypothetical method in which a suppositional theory is confirmed by observed effects. To be certain, differences appear between Cartesian and Aristotelian meteorology, as his opponents were sure to point out. Nevertheless, many key Aristotelian approaches to explanation and causation are found throughout Descartes' treatise without being changed at all or with only minor changes.

Assessing Descartes' familiarity with his contemporaries is difficult because of his tendency to deny that he was influenced by anyone. Nonetheless, we know that he was exposed to scholastic philosophy during and after his studies at the Jesuit college at La Flèche.[2] What impact his studies had on him is unclear. In a letter to Mersenne in 1640, three years after the publication of *Les Météores,* Descartes requested the names of Jesuit textbooks in philosophy because he could only remember those of the Coimbrans, Francisco Toletus, and Antonio Rubio.[3] Of those three, only the Coimbrans wrote on meteorology. In the same year he heaped fantastic praise on Eustachius a Sancto Paulo's *Summa philosophica* (a textbook on the entire range of philosophy, including meteorology), which he said he had just recently purchased. Descartes' description of his limited memory of earlier readings may very well be true, but it cannot be taken as proof of his total ignorance of contemporary Aristotelians in the 1630s. The correspondence between *Les Météores* and other Aristotelian meteorological works is evidence of at least a minimal familiarity with one or more of these books.

There is additional evidence that Descartes was acquainted with the content of recent treatises on meteorology. Soon after the publication of *Les Météores,* Libert Froidmont (1587–1653), a professor of theology and philosophy at Louvain and author of the well-circulated and frequently reprinted *Meteorologicorum libri sex,* first printed in 1627,[4] criticized Descartes on a number of grounds: his philosophy was too close to atomism, had unacceptable implications about the human soul, and could not adequately address the nature of sensible bodies. In a lengthy letter sent to Plempius and intended for Froidmont, Descartes responded to a number of Froidmont's complaints. With regard to *Les Météores,* Froidmont had written that Descartes' description of the composition of bodies by combinations of their parts and hooked shapes was "too crass and mechanical,"[5] and he complained that Descartes "hopes he will explain too many things by position and local motion, which cannot be understood without some real qualities"—real qualities, in Descartes' mind, being qualities that add a new reality to a substance rather than being merely a mode, such as locomotion or shape.[6] In sum, Descartes' meteorology suffered from its use of only matter and motion to explain complex phenomena. Descartes defended himself from Froidmont's critique not by arguing that "real qualities" were unnecessary or superfluous, a position he had taken in *Les Météores,* but by contending that his work compares favorably to those of his contemporaries. He wrote, "But if one should wish to list the problems which I explained only in the treatise *De meteoris,* and compare them with what has been done up until now by others on the same subject, in which he [Froidmont] is very versed, I am confident that he would not find such a great occasion for condemning my somewhat bloated and mechanical philosophy."[7] Perhaps he was bluffing about his knowledge of contemporary works on meteorology. Nevertheless, in his defense of himself, Descartes maintained that his meteorology addressed the questions typical of the state of the field, thereby suggesting he had some idea what the state of the field was and perhaps that he was aware that others relied on similar kinds of explanations, that they too were "crass and mechanical."

Indeed, a comparison of Les Météores with the Aristotelian meteorological tradition shows that Descartes was to a certain degree correct about a number of his Aristotelian contemporaries, if in fact his reference to mechanical philosophy was meant to imply the omission of final causes and substantial forms. Debates over the status of final and substantial forms in meteorological explanations had already begun well before Descartes and continued into the early years of the seventeenth century. The supposed novelty of eliminating substantial forms from meteorology, thus, was in fact no novelty at all in Descartes' times, since earlier Aristotelians from Albertus Magnus to John Poinsot had already defined the field as one that was overwhelmingly concerned with efficient and material causation. Descartes' meteorological theories should not be understood as an attempt to start an intellectual revolution but rather as a continuation of earlier debates. These debates were still ongoing in the 1630s, thus the publication of Les Météores prompted the responses, both hostile and more polite, of those who remained unconvinced that substantial forms had no place in explanations of meteorological phenomena or that the elimination of wonder should be a goal of meteorology.

Other points of similarity emerge from a comparison of Descartes' work to those of contemporary Aristotelians, such as Cabeo. While a number of followers of the Aristotelian tradition from antiquity until Cabeo's time understood meteorology as a field that primarily relied on material and efficient causation, Cabeo went further and used Aristotle's Meteorology as a model for all of natural philosophy. Aristotle and many of his followers recognized that the applicability of formal and final causes was limited in this field, but Cabeo and Descartes drew similar conclusions about all natural philosophy. They thought their discussions of the material and physical principles of meteorology served as a foundation for or an introduction to corpuscular explanations of natural phenomena in general. The kinds of explanations typically employed in meteorological explanation were based on the motion and combination of small particles and became the model for all of physics.

Although it seems unlikely that Descartes and Cabeo influenced each other, their approaches demonstrate that in some fields both Aristotelians and their critics were capable of arriving at similar conclusions. For Cabeo, Aristotle's *Meteorology*, with its emphasis on material and efficient causation, became the basis of a physics stripped of metaphysical aspirations, thereby retaining the prominence of Aristotelian texts in natural philosophy while greatly changing its direction. Descartes' *Les Météores* relied on similar tactics whereby his failure to employ substantial forms ideally would not provoke anger among schoolmen because the field traditionally utilized this kind of explanation only to a limited degree. He had suppressed publication of *Le Monde*, which treated the entirety of the natural world, in the early 1630s because he feared negative reactions from the Catholic Church. In contrast to the subjects of that work, meteorology had little bearing on topics of concern to the Catholic Church of the 1630s. The church was more concerned about topics such as psychology and cosmology. The same is true for the two other works, *La Dioptrique* and *La Géométrie*, which were published with the *Discours*. Descartes' reluctance to enter openly into the controversy in *Les Météores* is confirmed by his omission of comets, a topic that might well have led him into conflict with the orthodoxy of geocentricism. Therefore meteorology was an ideal topic for Descartes to develop his physics in a relatively noncontroversial manner.

LES MÉTÉORES

Les Météores was first published in 1637 together with *Discours de la méthode, La Géométrie,* and *La Dioptrique*. Descartes had contemplated and discussed many of the issues found in *Les Météores* throughout the late 1620s and the 1630s.[8] Some recent studies on Descartes' physics and *Les Météores* have emphasized the eighth discourse, in which Descartes explained the rainbow through a geometrical analysis of refraction.[9] This emphasis has given the appearance that his study of meteorology was part of Descartes' larger goal of applying mathematics to natural philosophy as he did in the accompanying *La Dioptrique*. Other recent scholarship, however, has emphasized the physical aspects of his account of the rainbow and how the deductive method Descartes used relied on observation and experience.[10] While it is clear that Descartes was especially proud of his treatment of the rainbow and saw it perhaps as emblematic of his new method,[11] much of *Les Météores* is not geometrical, depending instead on a corpuscular model that he explicitly presented as based on hypothetical or suppositional explanations. Moreover, while Descartes appeared to have been rightfully proud of his treatment of the rainbow, it should be kept in mind that this discourse is meant to be part of natural philosophy,[12] not mixed mathematics, and that it is much different from the previous seven discourses, which rely largely on descriptive accounts of the motion, shape, and size of corpuscles.[13] Furthermore, Descartes' discussion of the rainbow was not fresh by the time the *Discours* was printed. It appears that he had developed his solution to this question in 1629 and subsequently became interested in other meteorological problems in the middle of the 1630. Accordingly, Daniel Garber writes, "When the account of the rainbow appears eight years later in the *Meteors* it appears as a kind of ghost from an earlier period."[14] Concentrating on the problem of the rainbow distorts the focus of the entire treatise because most of *Les Météores* is not an attempt to ground meteorology on geometry but rather is an attempt to discuss how a wide range of atmospheric phenomena might be explained using only matter and local motion. This distortion becomes even more apparent if one looks at Descartes' writing on meteorology in the *Principia,* in which the accounts of exhalations composed of particles is maintained, whereas mathematical explanations of optical phenomena in the sky are noticeably less prominent or even absent.

Descartes intended the three essays that followed the *Discours* to illustrate his approach to natural philosophy and mathematics. In the letter to Father Dinet, he maintained that the method used in *La Dioptrique* and *Les Météores* was ideal.[15] The method used is hypothetical, not in that it mirrors contemporary views about scientific method but rather in the sense that starting assumptions about causes are taken to be hypothetical, not rigorously proven. Then these assumptions are shown to conform to experiential evidence and reasoning, thereby giving more confidence that these assumptions are in fact true.[16] In the *Discours* he described this method as one in which effects are deduced from hypothetical causes, and the experiential confirmation of these effects thereby explains their causes.[17] This method bears some similarity to the first part of the *regressus* method, which a number of Renaissance Aristotelians thought was the appropriate path toward understanding meteorology, based on their reading of *Meteorology* 1.7. Descartes was aware of this passage, if not in 1637 then at least by 1642. In *Les Météores* Descartes contended that the starting principles of meteorology were not proven, but their merits partially depended on their simplicity: "It is true that knowledge of these matters [i.e., meteors] depends on general principles of nature, which have not yet, as far as I know, been explained well, so it is necessary that I use at the beginning some suppositions [*suppositions*]. . . . I will try to render them so simple and easy that you perhaps will have no difficulty in believing them, even though I have not yet proved [*demonstrées*] them."[18] Thus Descartes made suppositions by which he believed he could give a clear explanation that is imminently believable even though they had not been subject to strictly defined proof.

Like many Renaissance Aristotelians, Descartes thought that at least some parts of his meteorology were conjectural because they were based on insufficient experiential knowledge, even if he thought that in the end he would be able to show the near certainty of the starting principles.[19] For example, after giving two potential explanations for why seventeenth-century sailors allegedly saw Saint Elmo's fire four or five times more often than ancient sailors did, Descartes justified his unwillingness to give a definitive theory by saying that "I cannot say anything but conjecture (*coniecture*) about what occurs in the great seas, because I have never seen them and I have only very incomplete accounts of them."[20] Thus the lack of access to direct observation rendered portions of his work conjectural. The offering of multiple explanations, a hallmark of uncertainty in Aristotle, Theophrastus, and Epicurus in antiquity, and in Averroes and Albertus Magnus in the Middle Ages, appears in other places in *Les Météores*. In a discussion of aerial apparitions (namely squadrons of battling ghosts), whose existence Descartes was skeptical of, he said that if they do exist then three distinct explanations might suffice for these purported phenomena. Either they are the result of a number of small clouds that emit thunder and lightning (small clouds that are struck by the lightning from a nearby storm) or these clouds are so high that they reflect the sun's rays.[21] No matter to what degree he thought his explanations of particular effects were only probable, Descartes presented his principles confidently. He believed that the principles he utilized—matter, size, and local motion—were simple, easily understood, and universally accepted. In a related fashion, he supposed that three kinds of corpuscles that differed in size and weight, despite their simplicity, could still eventually become the basis for complete accounts of sublunary change.

The simplicity of these principles lay partially in the lack of recourse to "substantial forms" or "real qualities," not, he wrote, because he denied their existence tout court but because he deemed them superfluous: "Then know also that in order not to break the peace with the Philosophers, I do not wish at all to deny that which they imagine in bodies beyond what I have said of them, such as their *substantial forms*, their *real qualities*, and similar things, but it seems that my explanations (*raisons*) must be approved so much more, since I will make them depend on fewer things."[22] Descartes' unwillingness to reject outright the existence of substantial forms and real qualities in *Les Météores* was likely a matter of delicacy. Years later, while advising Henricus Regius, Descartes suggested that this tactic was meant to illustrate that these concepts were of no use without incurring the anger of more-conservative Aristotelians. In a letter of 1642 he suggested that a diplomatic approach might allow Regius to avoid angering some of his colleagues at Utrecht, such as Gisbertus Voetius, who found Cartesian natural philosophy, with its rejection of Aristotelian concepts, plagued by potentially heretical and even atheistic ramifications.[23]

Les Météores presented a subtle path for Descartes' attempt to eliminate substantial forms from physics. This effort is widely regarded as central to Descartes' critique of scholastic natural philosophy.[24] Although Descartes seems to point out the principle of parsimony as the favorable criterion for not employing substantial forms and real qualities, he also disfavored them because he considered them unnatural to human thought and difficult to grasp. In the *Discours* while describing the principles he used in a writing that he had not yet published, namely *Le Monde,* he described how he had tried to avoid scholastic terms and concepts because of their relative lack of intelligibility: "And so first I described this matter [i.e., the matter that fills the universe], and endeavored to represent it in such a way that there is nothing in the world, as it seemed to me, clearer or more intelligible. . . . For I supposed, plainly, that it [the matter] possesses in it neither any of the forms or qualities, which are disputed in the schools, nor any other things the knowledge of which is not so natural to our souls, that no one can even pretend not to understand it."[25] Descartes was by no means the first to declare substantial forms unintelligible; it was a critique found in scholastic thought as well.

If substantial forms are incorporeal and not subject to direct observation, then substantial forms' roles as natural causes are at least slightly mysterious, since it is impossible to imagine a physical depiction of their actions. How they perform their functions or came to exist was also unclear, as scholastics debated how the qualities of substances emerged from substantial forms. For example, how does the heat of fire relate to the element's substantial form? Even before Descartes, some scholastics went so far as to declare that an understanding of substantial forms is beyond human capacity. Such was the view of Julius Caesar Scaliger, as found in his well-known *Exotericarum exercitationum,* a defense of Aristotelian natural philosophy first published in 1557. He wrote, "Such we see the narrowness of the human intellect: who would dare to say that he understands the forms of substances. Rather this exquisite cognition is hidden from us, just as how a unity is created out of two things. In what way is the form in the whole, and in the entirety, of each part? For what is that part of fire?

Therefore, form is a divine thing."[26] Thus for Scaliger substantial forms were divine, and, like the absolute power of God, they cannot be fully grasped by human knowledge.

Scaliger's discourse proved influential. Daniel Sennert cited the above passage in his 1636 *Hypomnemata physica*, in which he contended that natural philosophers could know nothing more about heat except that it "flowed from and depended on the form of fire," just as occult qualities arose from a form, which "was unknown by man."[27] In the following decade, Cabeo thought the physical elusiveness of substantial forms made them unsuited for natural philosophy. Whereas Cabeo decried traditional formulations of substantial forms for their lack of basis in experience, Descartes did so because they were "not so natural to our souls that no one can even pretend not to understand it."[28] Descartes had concluded that if substantial forms cannot truly be grasped, their etiological replacement should be clear. And that replacement was found in the size, movement, and position of corpuscles. Descartes' view that substantial forms were superfluous, unnatural, and complex continued to be prominent in his later works on natural philosophy.

WONDER

The intended simplicity and clarity of explanations had a moral element as well as an epistemological one, found namely in Descartes' aim to limit wonder at the natural world. Renaissance approaches to marvelous and wondrous meteorological phenomena varied. Some, particularly sixteenth-century Lutherans who rejected secondary causation,[29] maintained that what was wondrous was supernatural, directly emanating from God's will. By the middle of the seventeenth century, many Catholic thinkers were apt to be more restrained in their assessment of deviations from the ordinary. Cabeo, for example, ruled out appeals to divine action in physics, contending that such discussions were metaphysical, or at least not based on sensation, experience, and the physical world, while still conceding that God can make and has made miracles, such as the universal flood. The character of these miracles, however, fell outside the scope of natural philosophy. Similarly, the experts and intellectual elites who were summoned to discuss the purple rain that fell on Brussels in 1646 strove to find natural causes for this awe-inspiring precipitation and distanced themselves from what they saw as terrified or gullible uneducated masses.

Renaissance understandings of causation were tied not just to the theological but to the ethical realm as well. For example, Pomponazzi, despite his skepticism toward the possibility of complete human knowledge, followed a Stoic line of thought, contending that knowing the underlying causes of meteorology, as part of God's ordering of the universe, was the only possible hope for eliminating wonder and finding mental security and tranquility in a permanently tumultuous world.[30] Even though Descartes was not so ambitious in his *Les Météores* to suggest that an understanding of the causes of meteorological phenomena would bring about security, he nevertheless linked the grasping of causation to the elimination of the astonishment that accompanies ignorance.

Descartes' wish to check astonishment and explain the marvelous formed an important part of the scope of *Les Météores*; it forms the subject of the opening and closing paragraphs and reemerges in an explicit fashion throughout the work. His views toward wonder partially correspond to later lengthier treatments in the 1649 *Passions of the Soul*, which have become the touchstone for Descartes' treatment of wonder. There, Descartes distinguished between "wonder" (*admiration*) and "astonishment" (*estonnement*), the latter being a pernicious excess of the former. Although wonder potentially spurs investigations into nature, astonishment paralyzes the intellect and blocks further understanding.[31] These views toward wonder parallel those found in Isaac Beeckman's journals, which also included many discussions of meteorology. Descartes had spent much time with Beeckman in the years 1618 and 1619. In his journals, Beeckman wrote that "in philosophy one must always proceed from wonder to no wonder."[32] In *Les Météores*, Descartes followed the general outline of the treatment of wonder found in the *Passions* while not adhering strictly to the distinguishing nomenclature contained in that work. In the earlier *Les Météores*, wonder is the starting point of the investigation of meteorology and its elimination is the goal, just as it had been for Beeckman. "Naturally," he wrote, "we have more admiration (*admiration*) for the things above us than for those that are at our level or below." The upward gaze has led poets and painters to associate the clouds and skies with God's throne and imagine that God's hand "sprinkles dew on flowers and launches thunder bolts against cliffs." But these images of God's actions, however, are harmful fantasy. Thus Descartes intended to "explain their [i.e., meteorological phenomena's] nature in such a way that we should not have any more occasion to admire (*admirer*) . . . and we should readily believe that it is possible, in the same fashion, to discover the causes of everything that is most admirable (*admirable*) above the earth."[33] The final sentence of *Les Météores*, in which Descartes summed up his work, also reveals the perceived relation between causal accounts and the elimination of wonder: "I hope that those who have understood everything that was said in this treatise will never in the future see anything in the clouds for which they cannot understand easily the cause of and nothing that should cause them to wonder (*admiration*)."[34]

Because of his desire to eliminate his readers' wonder, Descartes was partially guided by awesome aspects of nature. The introductory discourse promises a later explanation of snowflakes "composed perfectly with six points," which is "one of the rarest marvels (*merveilles*) of nature."[35] He fulfilled this promise; the sixth discourse provides a lengthy examination of this topic. Similarly, Descartes prefaced his long account of the rainbow, of which he was especially proud, by describing the phenomenon as "a marvel (*merveille*) of nature so remarkable" that it has spurred learned men to look for its cause for all of recorded time.[36]

Descartes' explanations were intended to naturalize the marvelous, making it no longer the subject of astonishment. He wrote that tempests, lightning storms, and similar phenomena are "rather wondrous (*admirables*) for those who are ignorant of their cause" and then proceeded to provide his causal account, presumably alleviating the reader of wonder after the causes had been apprehended.[37] Similarly, he urged that we should not be astonished (*on ne s'estonne*) that the sun's light raises the exhalations to the upper levels of the sky because a careful consideration of the ordinary reveals that dust particles climb high above a lowly plain whenever numerous people trample across it.[38] Explanations thus eliminate the stupefaction concomitant with astonishment and indiscriminate belief in the miraculous. For example, not unlike many of his contemporaries, he believed that the diversity of particles within the exhalations explains why rain can differ in such extreme ways, giving a possible explanation for apparent prodigies, such as rains of iron, blood, or frogs.[39] Where Catholic authorities feared unauthorized attribution of miracles—such as in the case of the analysis of the purple rain in Brussels—Descartes sought to free his followers from unnecessary stupefaction by providing causal accounts of the same purported phenomena. Effective explanations required principles that were clear and did not produce stupefaction; substantial forms failed in this regard, at least according to Scaliger. For Descartes, these clear explanations derived from understanding that all epiphenomena stemmed from the position, magnitudes, and motion of small particles.

DESCARTES' CORPUSCLES

At the start of *Les Météores* Descartes followed the traditional distinction made in commentaries on Aristotle's *Meteorology* that substances can be classified as perfect or imperfect mixtures. Whereas in Aristotle's work, matter theory is not treated in depth until the fourth and final book, Descartes began his treatise with an exposition on the subject, making it the conceptual foundation for his exposition. He hypothesized that the traditional elements are composed of small irregular particles that join together, although never perfectly, leaving pores throughout the composition. Smaller particles that move more quickly than larger ones fill up any interstitial spaces created by the juxtaposition of these larger particles. These large particles move more slowly but possess more impetus and thus can agitate other particles easily. The motions, combinations, shapes, and positions of these particles give rise to the various secondary qualities of substances, such as the heat of fire or the coolness of ice, as well as cause their transformations and changes.

While rejecting traditional explanations of the elements, he retained the Aristotelian terminology of vapors and exhalations. For Descartes, vapors (*vapeurs*) are those bodies composed of fine materials that are present within the pores of terrestrial bodies. These vapors fill the microscopic spaces within other bodies, ensuring that the universe is a plenum. Exhalations (*exhalaisons*) are closely related to these vapors but are more regular in their shapes, being composed of particles with a shape similar to those that constitute water but finer. He likens exhalations to "spirits or eaux-de-vie," making an analogy to distillation.[40] Their fineness, that is, their small size, renders them more disposed to move quickly.

Even though Descartes' vapors and exhalations correspond greatly with Aristotelian conceptions, including their characterization as being made up of subtle matter, not all of *Les Météores* has such an obvious parallel in the work of contemporary Peripatetics. Perhaps the greatest novelty presented in *Les Météores* is Descartes' account of the sun's light. Light is composed of extremely fine particles. By colliding into the vapors and exhalations, the sun's light agitates them, thereby initiating the vapors' irregular but cyclical motion throughout the atmosphere.[41] These vapors and exhalations, which were ultimately composed of the same minute particles that form the entire physical world, are a constant resource in *Les Météores* and are the matter of winds, clouds, and lightning, among other things, just as they were for Aristotelians.

The understanding that corpuscles comprise the substrata of meteorological and other phenomena was common to Aristotle as well among Aristotelians.[42] Yet there were differences between Descartes and the Aristotelians as there were among the Aristotelians themselves. While Cabeo understood substantial forms to be a relation or bond between various corpuscles, a number of his predecessors contended that imperfect mixtures retained the substantial forms and qualities of the elements or of other mineralogical materials that are part of the terrestrial exhalation. Sennert and Froidmont believed they had experimental evidence that the vapors and exhalations retained the substantial forms of the elements, minerals, or chymicals that composed their underlying matter, which for them was bituminous, sulfuric, or nitric.[43] This contention was in turn useful because it could provide a conceptual underpinning for explanations of phenomena that were difficult or impossible to observe because of their location, far from the earth's surface. These scholastics argued that thunder and lightning behaved in similar ways to the terrestrial phenomena that was composed of the same matter. Descartes applied analogies from the terrestrial to the aerial as well but substituted the size of the particles for the qualities that derived from substantial forms. As a result, despite being based on a different conceptual foundation, many of Descartes' explanations are nearly interchangeable with Aristotelian ones. For example, just as many of his predecessors and contemporaries did, Descartes depended on chymistry. This interchangeability is particularly evident in his discussion of lightning and thunder.

Analogical reasoning constitutes many of the descriptions and arguments of *Les Météores*. Descartes' account of how the small corpuscles form macroscopic matter compares this process to branches being woven together and to eels that seemingly form a fluid substance in a container. Similarly, Descartes' explanation of lightning and thunder employed analogies to more easily observable phenomena. The starting point for his general theory for thunder is that it occurs when one cloud is situated above another. When the air surrounding the top cloud is hot, it condenses the vapor of the top cloud, thereby causing it to become heavier. The falling particles of the higher cloud produce a loud sound, namely thunder, when they collide with those of the lower cloud.[44] Descartes likened the production of thunder to the noise generated by an avalanche. He determined, based on his personal observations while travelling in the Alps, that the booming sounds of avalanches are the result of snow that, having been heated by the ambient air, becomes heavy and falls on the snow in the valleys below. The role that heat plays in this theory confirms everyday experience: "It follows that one can understand why it thunders more rarely in winter than in summer; for then not enough heat reaches the highest clouds, in order to break them up,"[45] although this heat, according to Descartes, is not to be thought of as a "real quality" but rather is equivalent to fast particles that agitate their quiescent neighbors.

Lightning is produced by the exhalations that are trapped between the two colliding clouds that thunder. A period of hot and dry weather will make the exhalation fine (*subtile*) and inflammable. When this happens the descending cloud is small, sometimes so small as to be invisible, but still produces small flashes. If the higher cloud, however, should be heavier and fall quickly it can produce thunder bolts if there is a heap of inflammable exhalations. Not deviating too greatly from the experience found in Aristotelian works, Descartes believed that these "fatty and oily" exhalations smell of sulfur. Other kinds of lightning are composed of such subtle and penetrating matter, similar in nature to salts or acids, that they can melt a sword without burning the sheath, an example found in nearly every Aristotelian discussion of lightning.[46] When the matter of lightning is a mixture of the penetrating kind and the fatty (*grasses*) and sulfuric (*ensouffrées*), they can produce a rock. Descartes believed that this possibility was demonstrated by an experiment not dissimilar to one offered by Sennert: "One can see through an experiment (*par experience*), that having mixed certain portions of earth, saltpeter, and sulfur, if one lights this composition on fire, it suddenly forms a rock."[47] It was widely held at this time that what we might recognize as prehistoric artifacts, such as arrowheads, were thunderstones.[48] While he did not employ "real qualities," his explanations include qualitative assessments nevertheless. These qualities (which are supposed to be reducible to descriptions of shape and size, such as fineness) fit with descriptions that Aristotelians employed as well. The condensing power of hot air, the inflammability of sulfur, and the penetrating fineness of acidlike exhalations render Descartes' account more vivid because the causes are related to sensible objects and the descriptions are in accordance with more common perceptions of the world.

CRITIQUES

Even if there are numerous similarities between Descartes' *Les Météores* and his predecessors' explanations of meteorological phenomena, and even if this work might be, in fact, "no more crass or mechanical" than the status quo, it provoked criticism. Two of the most sustained and significant attacks on the *Discours* and the *Essais* concentrated on *Les Météores*. Froidmont laid out eighteen points of contention, the last nine directed at *Les Météores*. Froidmont received a response from Descartes. The second attack was a vitriolic diatribe, written by a Hermetic chymist who used the pseudonym Mercurius Cosmopolita. That invective, which allegedly greatly annoyed Descartes, was personal as well as substantive, attacking Descartes' matter theory in addition to his perceived arrogance. Descartes made no response to Cosmopolita. These works illustrate how contemporaries found Descartes' *Les Météores* problematic because of its similarity to atomism and its unwillingness to address how the immaterial informs the material world, despite the text's similarities to some Aristotelian works.

Froidmont, himself the author of a meteorological treatise, as well as works on the soul, the immobility of the earth, and infinitesimals, was particularly worried by atomism, a philosophy that he believed threatened the Catholic faith.[49] This concern informed his critiques of Descartes, even though he apparently trusted Descartes' religious sincerity.[50] His letter to Plempius began with praise of Descartes' wit; he likened him to "some Pythagoras or Democritus, who as a willing exile from his country visited the Egyptians, the Brahmins, and the entire orb, in order to explore the nature of the universe." This praise, however, had a double meaning, since Democritus was the founder of atomism, and Pythagoras was the author of dubious propositions about the transmigration of souls. Despite Descartes' bright mind, according to Froidmont, "He receded into Epicurus's physics, crude and bloated, and not honed enough, as many believe, for a work of exact truth."[51] Froidmont himself utilized substantial forms and "real qualities" in his many works, including in his *Meteorologicorum*, where although he considered the *meteora* to be formed from imperfect mixtures, he believed that the substantial forms of water or earthy matter were preserved in the vapors and exhalations, even if they did not transform into new mixtures with distinguishable substantial forms.[52]

Froidmont's critiques of the *Discours* centered on issues related to substantial forms and the excessive use of material and corpuscular causes. It seemed unlikely to him that "such an ignoble and base cause" as heat, such as the heat that naturally arises in a dung heap, "could produce all the operations of the soul in the human body, except for the very actions of the rational soul." Similarly, if there is no difference between animals and machines, "what is the need to implant substantial souls in animals?" From this proposition, he saw the danger of impiety. He wrote that if similar causes are applied to the operations of the human intellect, then "perhaps the road to atheism is paved."[53] *La Dioptrique* was worrisome to Froidmont because in his view Descartes treated light as if it were a "flow of atoms," a proposition that is improbable because then light would be equivalent to the seemingly instantaneous local motion of corpuscles.[54]

Froidmont's suggestions for *Les Météores* are based largely on similar concerns about matter theory and the role of traditional concepts of natural philosophy, such as "real qualities" and substantial forms. The first critique questions whether collections of heterogeneous corpuscles could compose homeomerous bodies: "The composition, then, of bodies out of parts of different shapes (p. 159 [AT, 6:233]) by which they cohere to each other by hooks, seems crass and mechanical. For the many parts of water, for example, are uniform; and to the smallest parts of parts, it is not possible to distinguish such hooks and pins of different shapes. Therefore it must be admitted that there is an integral union between the proximate parts, between which there is no heterogeneity or inequality of shapes."[55] Thus Froidmont, echoing *De generatione et corruptione* 1.10, in which Aristotle defined a mixture as a union,[56] contended that a homeomerous substance, such as water, must be composed of unified proximate matter that is uniform and essentially akin to the substance and that the uniformity of the smallest sensible parts of substances such as water shows that each part is alike. Accordingly, he doubted that water was composed of a heap of long particles pointed in all directions, similar to eels in a bucket that are woven together like the fibers of animal flesh, as Descartes had maintained.[57]

Descartes' reliance on corpuscles could not adequately explain a range of other phenomena, according to Froidmont. The sensations of hot and cold do not derive solely from the impulses of particles at particular points because those sensations depend on qualities. Similarly, Froidmont was skeptical of Descartes' corpuscular account of the taste of salt, which he believed was caused by the penetration of particles into the tongue's pores: "He hopes that too many things will come about through only position, or local motion, which cannot [occur] without some real qualities, or I do not understand anything."[58] Furthermore, he dismissed Descartes' explanation of how the sun disperses and lifts the exhalations. Descartes' analogy to the dusty field being trampled is unhelpful because it assumes that the sun's light is corporeal, the same criticism that he had sustained against *La Dioptrique*.[59] Furthermore, Froidmont questioned Descartes' use of motion to define some qualities: the fineness of vapors could not be caused by the quickness of motion because still bodies are just as capable of being fine as moving ones.[60]

While Froidmont disagreed not only about the extent to which corpuscles, position, and motion could explain meteorological phenomena but also about whether "real qualities" and substantial forms exist, the differences between his work and Descartes' were not extreme. Froidmont did not dispute that corpuscles formed some meteorological phenomena. These corpuscles, however, have substantial forms that explain their powers and the qualities of the vapors and exhalations, qualities, such as the fineness that Descartes admitted to be present in vapors. Descartes considered these qualities not as primary qualities but rather as epiphenomena that emerged from the arrangement of corpuscles of varying shapes and sizes. In a similar vein, Froidmont held that the sun caused vapors and exhalations to rise but that this was not done through particles of light but through the sun's natural power and heat; heat was a quality with its own powers and was reducible to more than just motion. Froidmont described the general material and efficient causes of *meteora* as follows: "There is no characteristic feature to the exhalations other than a light substance which is loosened from heavy bodies by the rarefying power of heat."[61] Heat makes bodies fine, which then causes them to rise in the form of vapors. The sun's power is visually evident, according to Froidmont: "We discern with the eyes that earth, when wet with rain, steams, [and that] the dry cracking of fields, struck by the sun, emits something fine like a fiery breath."[62] Despite not believing that light is material, Froidmont employed vocabulary that resonated with Descartes' words: the sun rarefies, and exhalations are fine or even *subtilissimum*, just as Descartes' primary parts of matter can be *subtile*, meaning small in size.[63]

Descartes responded to Froidmont's critique by embracing the epithet *mechanical* and by denying that he was an atomist, a denial that Descartes thought was unnecessary due to his rejection of voids, one of the two primary principles of both Democritean and Epicurean atomism. "For how could that [epithet] pertain to me, when in no place do I suppose a vacuum, but in contrast I say expressly that all space . . . is full?"[64] While it is unclear precisely what Froidmont meant by *mechanical*, Descartes wrote that "if my philosophy seems crass to him, so that he considers the shapes, magnitudes, and motions as Mechanics, then he condemns all that I value as praiseworthy."[65]

Describing mechanics as a field that did not rely on difficult metaphysical concepts but rather those simple principles that Descartes so valued, he promoted the subject as "the truer and less corrupt part of philosophy."[66] His conception of the mechanical seems defined partly in a negative way as well; it is an examination of the natural without recourse to "real qualities." Immediately after quoting Froidmont's statement that he cannot imagine successfully explaining the taste of salt by only position and local motion without "real qualities," Descartes contended that his *Les Météores* does not differ too greatly from other works on the meteorology and that Froidmont "would not have the occasion for damning my philosophy as bloated and mechanical" if he had compared it to other works on meteorology.[67] Thus Descartes must have assessed this field as one in which discussions depended on local motion, magnitude, and position to a greater degree than other parts of natural philosophy, thereby making it an ideal topic for introducing his method in a manner palatable to those accustomed to Aristotelian philosophy.

Daniel Garber has argued that while the *Discours* was attacked, it was not perceived as revolutionary by conservative Aristotelians, such as Froidmont.[68] It does not appear that Descartes perceived it as revolutionary, despite his recognition of some of its novelty; he did not appear to expect anything but widespread acceptance of the contents of *Les Météores* among those teaching in Jesuit colleges.[69] Although Descartes suppressed *Le Monde*, *Les Météores*, due to its focus on the inanimate terrestrial world, was a less controversial vehicle to present his larger goal of devising a physics that had no recourse to formal substantial forms. Furthermore, because the subject matter of meteorology did not demand discourses on cosmology, which had proved dangerous to Galileo among others, and because a certain camp of Aristotelian natural philosophers—which included not just Cabeo but also Poinsot and Albertus Magnus—was already in agreement with his reliance on efficient and material causation, *Les Météores* was a less dangerous vehicle than *Le Monde* to provide a new model for natural philosophy.

THE HERMETIC RESPONSE

The second response to *Les Météores* is found in a dialogue, or as the author called it, a pentalogue, with the title *Pentalogos in libris cujusdam Gallico idiomate evulgati quatuor discursuum, de la method; dioptrique; météorique; & géometrique*. The work was printed under the obvious pseudonym Mercurius Cosmopolita. Despite its title, the work is almost, if not completely, dedicated to attacking *Les Météores*, a fact that was not lost on its readers; Samuel Hartlib referred to it as the "Treatise Contra Meteora Cartes."[70] Erik-Jan Bos has convincingly identified the author as Andreas Haberweschel von Habernfeld, a Bohemian chymist, who at the time was living in The Hague. The well-travelled Habernfeld authored several works on Hermetic philosophy, medicine, and prophetic politics, in addition to being involved in a plot against the king of England and Archbishop Laud in 1640, the same year in which the *Pentalogos* was printed.[71]

The names of the five interlocutors who appear in the *Pentalogos* give a major clue to the nature of the entire work. They are the son of Hermes; the grandson of Apollo; Nature; Mercury, the son of Nature; and, the Cartesian figure, the Vainglorious Naturalist (*Naturalista gloriosus*). Despite the five characters, the work is more or less a dialogue between Mercury and the Vainglorious Naturalist. The dialogue, despite containing specific complaints about the content of *Les Météores,* is largely an ad hominem attack, which apparently irked Descartes.[72] The *Pentalogos* portrayed him as not just vain but also insecure, arrogant, and proud. The initial speech of the Naturalist is an obvious satire of the *Discours:*

> Good day, Sirs, good day. Oh I see that you turn through my pages, how does this book seem to you? Do you find to your liking the book, which declares to you the principles of things and enumerates the causes? I am the author of it. Am I not a learned man? What do you think of me? I frequented many schools, but I consumed there nothing but mistakes; for they are schools of mistakes. But nevertheless, having followed my genius, I touched upon the path to truth. I have contempt for these books, in which all types of knowledge are expressed universally, because they are far from the truth. I girded myself, and I assigned my genius to it, so that I would study in that great book of the world; and having ignored the view of others, I decided my own judgment is sufficient.[73]

With this mockery, quite unlike Froidmont's relatively polite tone, Cosmopolita offered some critiques similar to those of the professor at Louvain.

In this work, the Vainglorious Naturalist endorses atomism, equating atoms to the minute particles that compose earth, water, and air. The character Mercury, however, cannot accept that these atoms are sufficient to explain meteorological phenomena. More principles are needed than position, magnitude, and motion. According to Mercury, "Everything is of a double form, spiritual and corporeal, or material and formal, as there are material elements, and formal elements, material principles, formal principles, material seeds, and formal seeds, where the set times of predestination await."[74] The endorsement of Hermetic principles, which partially correspond to traditional Aristotelian distinctions between matter and form, accompanies an endorsement of Hermetic chymistry. Disputing the Cartesian position that vapors crashing into each other can cause thunder and lightning, the interlocutor Mercury contended that in addition to the aqueous vapors, "terrestrial, sulfurous, mercurial, nitric, antimonic, arsenic, and sandaracal exhalations" cause a variety of storms. Thunder itself comes from the exhalations of mercury and sulfur, while lightning emerges from the interaction of an exhalation formed from niter-mercury (with its icy spirit) and sulfur (with its "powerful fiery spirit").[75]

Despite their polar intellectual orientations, Cosmopolita found some sympathy with Descartes; they both shared a distaste for academic curricula. While Descartes lamented the reliance on what he considered unclear concepts derived from medieval authorities, Cosmopolita wished that universities would incorporate Hermetic principles into the teaching of chymistry. Even at Leiden, where chymistry had already become part of the training for the production of medicines, the university ignored the true principles, according to Cosmopolita: "There they will eventually fall into errors: these courses are not chymistry, but hairdressing, pseudochymistry, imposture, and delirium. They have not yet added the true chymists, students of *Hermetic Medicine*; for chymistry is the handmaiden of the Hermetics" (emphasis in the original).[76] Moreover, he believed that the schools' emphasis on anatomy debased natural bodies, which should be admired as part of the universe that God wove out of the fabric of chaos.[77]

The author's positive orientation toward Hermetic philosophy underscores the biggest gaps between him and Descartes. Descartes' reliance on matter and motion to explain meteorology ignores what Cosmopolita thought were the central truths of the universe, that its creation was the result of immaterial spirits that continue to shape the constituents of the world, that air is not just an element and a spirit but is also the "world soul" (*spiritus mundi*). For the Hermetics, nature is a divine spirit, and thus products of nature must be wondrous. For example, the causes of the colors of the rainbow, which are a sign of the pact between God and humans, should be admired rather than analyzed, according to Cosmopolita.[78] The contrast between Hermetic philosophy and Descartes is emphasized by the Vainglorious Naturalist's assertions that wonder is the result of human weakness and ignorance, which his work will remedy.[79] Even though this portrayal of Descartes is perhaps an overstatement, it reflects his intention that *Les Météores* would remove his readers' stupefying wonder. Descartes' attempt to remove wonder by employing what he thought were intelligible principles was not lost on his contemporaries.

METEOROLOGY IN THE *PRINCIPIA PHILOSOPHIAE*

The meaning and intent of *Les Météores* is revealed in part in Descartes' subsequent treatment of meteorology. His own reaction to and assessment of *Les Météores* and its method were ambiguous in the 1640s. On the one hand he maintained that he employed the true principles in this work, as in his 1642 *Letter to Dinet*; on the other hand, he pointed out some of the rhetorical strategies that would help his philosophy gain greater acceptance. In the 1642 *Letter to Regius*, for example, he explained that he did not reject outright substantial forms and "real qualities" so that he would not be required to explain such a rejection. Ideally the reader would see how convincing his explanations were without recourse to these concepts; the reader would independently come to the conclusion that they should not be part of explanations within natural philosophy. Descartes was also aware of the similarities between his hypothetical method and Aristotle's description of method as described in the *Meteorology*. These similarities, Descartes believed, would aid his defense from attacks that his philosophy was a dangerous novelty, which was the view of Gisbertus Voetius.

Voetius was a member of the theology faculty at the University of Utrecht. He was a staunch Aristotelian and maintained, in the words of Theo Verbeek, "that everything theology needs happens to be found in Aristotle's work." [80] His hostility toward Descartes' philosophy emerged in 1641, when the physician and professor of medicine at Utrecht, Henricus Regius, an early convert to Cartesian thought, after having read the *Discours* and the *Essais*, declared that "homo est ens per accidens" (man is a being by accident). [81] Voetius considered this statement to be heretical because it implicitly denied that the human soul is man's essence. While Descartes did not endorse Regius's statement, it exemplified the possible heretical ramifications of the rejection of substantial forms, which authors like Froidmont had feared as early as 1637. At any rate, Voetius's critiques and Descartes' replies soon took on a vitriolic character, which ended in the declaration of the city government that Descartes' philosophy could neither be taught nor criticized publicly. In the course of these exchanges, Descartes realized that he could make his natural philosophy appear more acceptable by demonstrating its correspondence to Aristotle.

In Voetius's appendix to his *Corollaries*, which was primarily concerned with the absence of substantial forms in Descartes' philosophy and Regius's reformulation of it, Voetius complained that Descartes had "neither explained nor demonstrated any of the mysteries of nature (*in naturae arcanis*)." [82] Descartes responded to this critique of his hypothetical method by answering in similar way to his reply in his earlier exchange with Froidmont: "Those who say that nothing is explained through these principles should read our *Meteora* and compare it with Aristotle's *Meteora*." [83] What Descartes meant by this vague statement becomes clearer from what he wrote in a letter to Father Vatier in early 1638. There, in response to criticisms that he had not proved anything, he wrote that he considered his proofs not to be deductive or a priori but rather a posteriori, that is, the experiences demonstrated the principles. Aristotelians, versed in the regressus method, considered the term *a posteriori* as equivalent to the first stage of the *regressus*, the stage that was utilized above all others in meteorological investigations. [84] The links between Descartes' method and those of Aristotelians and Aristotle himself emerge in greater detail in the *Principia philosophiae*.

In the *Principia,* first printed in 1644, Descartes used a matter theory consistent with that of *Les Météores* to underpin his explanations of the natural world, which he argued was composed of three elements of varying size. The explanation of meteorological phenomena is largely identical to those in *Les Météores,* which Descartes cited several times in the *Principia.*[85] These three elements were the starting point of his hypothetical method as applied to the natural world in both *Les Météores* and *Principia.* Yet Descartes' assessment of his method had slightly changed since 1637, when he believed that the simplicity and clarity of the starting principles would guarantee their certainty or near certainty. To the contrary, at the beginning of the fourth part of the *Principia,* where he gave an account for the earth as well as for meteorological phenomena, Descartes described his method as one in which after making his hypotheses he would show that "the causes of all of the natural world" could happen "in this way but not in any other,"[86] making his account at least morally certain.[87]

Anticipating objections from traditionalists such as Voetius, Descartes contended that Aristotle used this method as well to explain phenomena that were inaccessible to accurate observation. Citing the same passage that Nifo used to support his claim for the conjectural nature of physical science,[88] Descartes defended his hypothetical method, which he likened to surmising the unseen internal mechanisms of clocks. Two clocks can have different internal states that in turn cause the clocks' hands to point to the same hour and minute. In his support of this position, he cited Aristotle: "And in the case anyone happens to be convinced that Aristotle achieved—or wanted to achieve—any more than this, he himself expressly asserts in the first book of the *Meteorology,* at the beginning of chapter seven, that when dealing with things not manifest to the senses, he reckons he has provided adequate reasons and demonstrations if he can simply show that such things are capable of occurring in accordance with his explanations."[89] Here, Descartes referred to the passage in which Aristotle wrote, "We consider that we have given a sufficiently rational explanation of things inaccessible to observation by our senses if we have produced a theory that is possible."[90] Whereas a number of Renaissance Aristotelians took this statement as an endorsement of the standard of "saving the appearances" for meteorology, Descartes thought this method was more than instrumental and could bring about moral certainty, if the result could give an explanation of all known phenomena. That is, he believed his theory was certain "as far as is sufficient for its application in life," though uncertain in respect "to the absolute power of God."[91]

Descartes saw the advantage of linking his method with Aristotle's text for future skirmishes with Voetius or others. In a letter, written in 1644 to an unknown recipient, he thanked his correspondent for pointing out the similarities between his method and that found in Aristotle's *Meteorology*. He wrote, "I was happy to see in the letter, by which you honored me merely by writing it, that you advise me to look at the beginning of the 7th chapter of the first book of the *Meteors* of Aristotle, in order to serve my defense. For it is one place that I have cited at the end of my philosophy and the only Aristotle that I have cited. Because it is not a small proof of your affection to see that you rightly advise to me the same thing that I believed must serve me." [92] The only citation of Aristotle in Descartes' works published in his lifetime is perhaps surprisingly not an attack but a defensive move designed to show the similarities between Peripateticism and his own philosophy, which would help defuse charges that his philosophy was dangerous.

—◌ ◌—

Descartes' response to Froidmont's critique of *Les Météores* for its closeness to atomism and its similarity to mechanics emphasized that his work was not much different in these respects from his contemporaries and predecessors and that the kinds of explanations used by at least some Aristotelians were similar to those Descartes employed. By the 1640s when he anticipated attacks on his method, his response shifted from asserting a similarity to his contemporaries to asserting a similarity to Aristotle himself. Potentially, this tactic was rhetorically useful against figures such as Voetius, who believed the fundamentals of human knowledge were contained in Aristotle's writings. Descartes' citation of Aristotle also reveals the multiple ways of interpreting Aristotle's text. Just as Cabeo used the *Meteorology* in order to fuse chymical theory and practice with Aristotelianism, paradoxically Descartes had found an authority for his own hypothetical method, the origins of which derived from a rejection of all authorities. The links between Descartes' thought and Aristotelianism was not lost on his contemporaries. Citing the example of Henricus Reneri, who gave public lectures on the *Discours* and the *Essais* in 1638 and made no attempt to address the thorny issues in Descartes' metaphysics, Verbeek concludes that "the conceptual barrier that prevented Dutch Aristotelians from embracing Descartes' physics was small." [93]

The conceptual barrier was small in part because Descartes most likely developed his meteorological theories while contemplating what was found in contemporary meteorological treatises. The correspondences go beyond similarities in content and method. They extend to specific explanations as well. His contention that the matter of lightning is fatty and sulfurous, his use of vapors and exhalations, and his display of knowledge about imperfect and perfect mixtures demonstrate that he developed his meteorological theories while considering the solutions of his contemporaries. Influenced by Aristotelians, he appropriated from their works and formed his own explanations with them in mind. The originality of Descartes' theories was not absolute; his ideas about meteorology had a basis in traditional natural philosophy even if some traditionalists rejected them. Yet Aristotelian understandings of meteorology bolstered Descartes' ability to reject substantial forms and support a suppositional method for terrestrial physics.

The connections between the meteorology of Aristotelians and *Les Météores* are indicative of the role that Peripateticism played in transforming natural philosophy. Much of the significance of Cartesian thought was in its proposing a path for understanding natural processes primarily based on the position and motion of particles. Stephen Gaukroger has argued that these explanations were central to the development of what would become known as "the mechanical philosophy."[94] Whether or not one accepts the appropriateness of that analysis, it is clear that Descartes greatly influenced natural philosophy as it was practiced in the second half of the seventeenth century, particularly in France and other intellectual centers on the Continent. The similarities between the ideas of Cabeo and Descartes, as well as the practices of subsequent authors who tried to synthesize Aristotelian and Cartesian natural philosophy, suggest that, partly because of considerations of the *Meteorology,* corpuscular explanations were palatable, if not preferable, to both Aristotelians and their opponents. Although the suitability of the term *scientific revolution* is debated in current scholarship,[95] large changes in the intellectual foundations of natural philosophy undeniably took place during the seventeenth century. The transformations of Aristotelian meteorology were part of those changes.

Epilogue

The emergence of modern science during the seventeenth century traditionally has been portrayed as the result of both the demise of Aristotelianism and
its replacement by innovative forms of natural philosophy that privileged
mathematical, experimental, or mechanical methods rather than textual explication and metaphysical concepts derived from ancient authority. This view
is found not only in recent scholarship but in the writings of the *novatores* themselves. Galileo and Bacon saw Aristotelianism as either a form of pedantry
or idolatry. Echoes of Galileo's and Bacon's denunciations that have rattled
throughout twentieth-century studies, however, stem from a limited engage-
ment with Renaissance Aristotelian natural philosophy. What is equated with
Aristotelianism is typically syllogistic argumentation or metaphysical concepts, such as matter, form and privation, potency and act, or the cosmological
schema—geocentricism—that increasingly were seen to fail to conform to observable phenomena. Thus emphasis has been placed largely on metaphysics
and the principles of natural philosophy in assessments of Renaissance Peripateticism. While no doubt important, examinations of these aspects of late Aristotelianism have skewed perceptions of the nature of Aristotelianism, which
encompassed not only debates about the principles but also the application of
them.

Because Aristotle in the *Meteorology* was more concerned with explanations
than with establishing principles (though he was concerned with establishing
principles in *Physics*, *De caelo*, and *De generatione et corruptione*), an understanding of meteorology is an important corrective to portrayals of Aristotelianism

as overly conceptual and bookish. Renaissance meteorology exemplifies the practice of natural philosophy with regard to nature rather than the principles of nature and with regard to particulars rather than general overviews. The history of Renaissance meteorology offers a counterpoint to the perception of clean or revolutionary breaks from Aristotelianism during the early modern period. Renaissance meteorology demonstrates not just the flexibility of Aristotelianism and the tendency of Aristotelians to see their theories as provisional but also internal concerns with the applicability of final and formal causation, the willingness to adapt theory to developments derived from empirical undertakings, such as those conducted in chymistry, and the adoption of corpuscular motifs as bases for material explanation. Additionally the philosophical inquiries into the teleology, causation, and epistemology of meteorological phenomena were not intellectual activities of those isolated within universities. The views of professors had ramifications in practical arenas, despite later scholars who criticized the methods of the schools. While now the practical is typically identified with the technological, in the sixteenth century the conclusions of learned meteorologists more often found a role in political and religious controversies. The Ferrarese earthquakes and the analysis of the purple rain in Brussels demonstrate that religious and political authorities deemed the conclusions of Aristotelians suitable for controlling public opinion and maintaining civil authority.

Opponents of Aristotelianism have frequently described it as unreceptive to new ideas and evidence. Yet developments within Aristotelian meteorology paralleled ideas endorsed by creators of natural philosophies that actively opposed the intellectual products of universities. Descartes was aware of these parallels and tried to profit from them, using meteorology as an innocuous way to introduce his approach to natural philosophy that did away with formal causation and based itself on the position, size, and motion of small particles. That Descartes' *Les Météores* and the pertinent sections of the *Principia* offer a version of meteorology close to Aristotelian ones should not be taken as out of the ordinary. In fact vapors and exhalations continued to be the primary mode of explanation throughout the seventeenth century.

The perception of similarities between Descartes' and Aristotelians' works is evident in attempts to combine the two. One such attempt was made by Jean Baptiste du Hamel, whose *De meteoris et fossilibus libri duo* (1660) applied chymical concepts in his explanation of meteorology and the formation of minerals. Du Hamel contended that the motions of vapors, smoke, and exhalations were the underlying causes of meteorology. These three categories of matter were defined by their shapes. Using terminology reminiscent of Descartes' evocation of eels and branches in *Les Météores*, Du Hamel described vapors as being composed of "long and slippery particles" and smoke and exhalations as being composed of "twiggy and irregular parts." Their shapes, however, were not sufficient to explain all of their properties, as Du Hamel invoked explanations appropriated from Aristotelians. Vapor emerges from water, smoke from the earth and "oily juices," and exhalations from "[even] more bitter juices."[1] Du Hamel followed Descartes in his elimination of formal causes, yet his definition of the subject of meteorology still depended on the distinction between perfect and imperfect mixtures. As a result, his definition of imperfect mixtures is akin to Cabeo's understanding of the physicality of substantial forms. Imperfect mixtures are those whose parts "can be separated easily."[2]

An approach similar to Du Hamel's is found in the work of Jacques Rohault, whose comprehensive treatment of natural philosophy was translated into both Latin and English, printed in many editions, and utilized as a standard introductory textbook throughout Western Europe. Following Descartes, Rohault failed to pronounce on the existence of substantial forms and consciously avoided applying them in his 1671 *Traité de physique*.[3] Despite his obvious Cartesian slant, it is difficult to distinguish his account of meteorology from Aristotelians' accounts, as Rohault utilized "la vapeur & les exhalaisons" as the material causes of meteorological phenomena in general, and "exhalaisons soufreuses, & d'autres exhalaisons plus terrestres" for his explanation of lightning.[4] The proviso that concludes the third book of Rohault's treatise, the book that treats meteorology, also hides its inspiration. It could be either Cartesian or Aristotelian. Rohault wrote that "no mortal can explain everything" and that as a result this third book is not truly "finished."[5] The magnitude and diversity of the field rendered his conclusions provisional.

The contention that knowledge of meteorology was incomplete or uncertain was widespread in the middle of the seventeenth century, common not just to Cartesians and Aristotelians. Teleological knowledge continued to prove to be a domain of uncertainty even for promoters of the mechanical philosophy, not all of whom had altogether banished final causation from natural philosophy. Meteorology provided a case in point for Robert Boyle. Despite admitting that the final causes of the sublunary world were the "most conducive to the universal ends of creation, and the good of the whole world,"[6] he was nevertheless hesitant to make unequivocal conclusions about the nature of teleology for meteorology. After considering the final causes for the celestial bodies, he then wrote, "That 'tis yet more unsafe, to ground Arguments of the Nature of particular Bodies that are Inanimate in the Sublunary World, upon the Uses we think they were design'd for."[7] Thus Boyle's views follow along the lines of Pomponazzi's, although without any recourse to Thomistic theology. For both, meteorological events preserve the good of the world, yet exactly how they do so is beyond the realm of natural philosophy.

During the latter half of the seventeenth century, the resonances of Renaissance uncertainty extended from teleology to the nature of theory in general. Joseph Glanvill, in his 1668 *Plus ultra; or, the Progress and Advancement of Knowledge since the Days of Aristotle*, noted that the barometer is a useful tool in measuring the weight of air, yet he conceded that theories that resulted from its utilization were likely eventually to be discarded, just as portions of Aristotelian theory had been. He wrote, "For we have no reason to believe it should have *better luck* than the *Doctrine* of the *Circulation*, [and] the *Theory* of the *Antipodes*" (emphasis in the original).[8] Recent development and instruments might provide new theories, which in turn were likely to be as provisional as Aristotelian ones had been.

Glanvill's mitigated skepticism saw improvements on Aristotelianism partially in the discovery of new facts that contradicted the statements found in Aristotle's works. One such fact that was long related to the meteorological tradition was what Glanvill called the "*Habitableness* of the *torrid zone*," which was one of "the most *general* and *necessary* things," unknown to Aristotle.[9] Glanvill was referring to Aristotle's treatment of the earth's climates in *Meteorology* 2.5, wherein he argued that the terrestrial globe should be divided into five belts or climates that wrapped around the world's circumference. Assuming that the borders of the climates must reflect the astronomical divisions of the earth, Aristotle defined the borders by the two tropics and the Arctic and Antarctic circles. Two of these zones were thought to be habitable and were called *oikoumenai*; one lies in the north, between the Tropic of Cancer and the Arctic Circle, and a corresponding region lies in the southern hemisphere. Between the tropics, around the equator, lies an uninhabitable torrid zone, too hot for the streams and pastures that are necessary for human life.[10]

Although Glanvill pointed to Francis Bacon in this section as the one who had demonstrated that the "*Ancients* were exceedingly *defective*," numerous if not all sixteenth-century Aristotelians had already rejected the torrid zone because of reports from sailors.[11] Agostino Nifo did so as early as the 1520s, contending that new experiences overrode the conjectural theories that were put forth by Aristotle. Nifo wrote that perhaps in Aristotle's time there were no reports that contradicted his hypothesis. By Nifo's time, however, there were such accounts, "thus what Aristotle attempted to establish by conjecture, is [not] verified by history."[12]

This passage from Nifo's commentary was not extraordinary; many Aristotelians asserted the primacy of experience over not only ancient authority but reason itself. The Coimbrans, who were at times characterized as among the most reluctant of Aristotelians to integrate new material in their commentaries, cited Christopher Columbus by name and argued that in this case experience demanded a reformulation of the earth's climates.[13] Much earlier, Pomponazzi considered the need to overturn reasoning because of experience as an essential part of natural philosophy. For him experience held a higher position than theory. Citing Avicenna's *Canon* and Aristotle's *Physics*, he wrote that "when reason is opposed to experience, then reason must be thrown out and experience must remain."[14] Thus, while Aristotle at times used an insufficient amount of evidence in induction, Pomponazzi nevertheless saw Aristotle as privileging experience over reason in explaining natural phenomenon. After asking why Aristotle put forth only the phenomena without giving an explanation when he described how honey is affected by moisture and the cold, Pomponazzi answered with the dictum that "it is possible to be known better through experience than reasoning."[15]

The widespread conviction among Aristotelians that experience should alter and improve upon reasoning and theory was key to developments in Aristotelian meteorology. Observations and investigations of particular events, such as the earthquakes of Ferrara during the 1570s, pointed to the untenable nature of Aristotelian natural philosophy for some, but to others it merely provided an opportunity to refine and adjust accepted explanations. As the range and kinds of experiences available changed during the Renaissance, Aristotelians integrated new methods that arose from chymistry and balneology into their theories, not just utilizing corpuscular explanations but also employing tests to help determine the underlying nature of the exhalations and to determine questions regarding the presence and persistence of substantial forms.

In the seventeenth century, a significant change in the field of meteorology was the rejection of the radical division between sublunary and supralunary. Some of the strongest critiques of this bifurcation came from proponents of novel natural philosophies. For example, Kepler argued for common causal explanations for the celestial and meteorological, Descartes contended that the rules of nature prevailed universally, and Galileo used observations taken with telescopes to argue for the mutability of the heavens. Some Aristotelians opposed these innovations. Cesare Cremonini allegedly refused to look

through Galileo's telescope because it provoked headaches, and Scipione Chiaramonti vehemently argued against the Tychonic and Copernican systems. Instead, he proposed alternative interpretations of recent supernovae and maintained that comets were sublunary.[16] Other Aristotelians, however, were more
open to changes. Daniel Sennert believed some comets to be supralunary, as
did Libert Froidmont.[17] The Jesuit Cabeo contended that comets were celestial
and that the mutability of the heavens demonstrated that Thomas Aquinas's interpretation of the hylomorphic nature of the substance of the orbs was correct.[18] Thus in discussions of this topic, Aristotelians displayed heterogeneous
views, both adapting to novelties in natural philosophy and reconciling these
novelties with past authorities such as Thomas.

The eventual rejection of Aristotelian and Ptolemaic cosmology, however,
did not dramatically shift the explanations of meteorology. The application
of vapors and exhalations to explain the material basis of meteorology continued throughout the seventeenth century and into the eighteenth, when
electrical and new chemical theories replaced the exhalations. Aristotelian
meteorology did not suffer a sudden death. Rather it persisted because of its
provisional nature, its concern with the material, and the success and flexibility of its primary forms of explanation, while portions of it continued to be integrated into newer systems despite the crepuscular character of early modern
Peripateticism.

The heterogeneity of Aristotelianism arose from not only the various positions of particular professors but also the diversity of the Aristotelian corpus as
a whole. Aristotle's writings differ not just in respect to the subject but in respect to the approach as well. Aristotle was well aware of this and even pointed
out that the kinds of knowledge and proofs should reflect the objects of inquiry.
As a result his treatises show different aspects of his thought. Because the *Meteorology* and its commentary tradition have been much neglected by modern
scholarship, the kinds of explanations and considerations of experience have
largely been left out of considerations of natural philosophy, as scholars have
concentrated on looking at logical works and commentaries on the *Metaphysics*
and *Physics* in order to explain Aristotelianism and its views of nature. Commentaries on these works often correspond to the critiques of seventeenth-
century *novatores*: they are filled with concerns over syllogisms and metaphysical distinctions while often devoid of considerations of experience much less
experiment. But the title of the *Physics* should not mislead. The work is largely
dedicated to the general principles of natural philosophy, and as a result it is
concerned more with the metaphysics of change, time, and space than with

investigations of natural particulars. The *Physics'* concern with metaphysical topics does not mean that Aristotle or his followers were not concerned with other matters in natural philosophy. Thus if we wish to see how Aristotelians pursued inquiries into nature, as opposed to inquiries into the principles of nature, the *Meteorology* and later meteorological writings provide the appropriate window. The *Meteorology* and its tradition offer insight into the practice of natural philosophy rather than the establishment of the principles of the natural philosophy. While there are some correspondences between the content of logical works and the approaches found in meteorological investigations, it is a mistake to assume that the idealized methods found in the *Analytics* that promoted syllogistic argumentation were applied uniformly to investigations of meteorology or other subjects of natural philosophy.

Once the division between idealized Aristotelian method and natural philosophical practice is admitted, the characterization of Aristotelianism as constricted by its demand for syllogism and logical rigor disappears. In this new light, we can see Renaissance works on meteorology as open inquiries, which attempted to construct theories that corresponded to sensible evidence, observations, ancient texts, religious doctrines, and experiment. While the explanatory core of exhalations and vapors remained beyond the life of Aristotelianism, understandings of them changed in accordance with new technologies and experimental techniques. The depiction of Aristotelians as unchanging, devoted to a system rather than to truth, does not hold for Renaissance authors on meteorology. The willingness of Cabeo to reject the use of immaterial substantial forms in natural philosophy, preferring to think of form as material links or chains between particles, illustrates the potentially extreme position of Aristotelianism, as Cabeo used interpretations of the *Meteorology* as a means for undoing, or reinventing, one of the most fundamental metaphysical entities of medieval and Renaissance thought.

The similarities between Aristotelians and some *novatores* extend beyond the issue of explanation and touch upon the goal of natural philosophy. The purpose of meteorology was in part moral. Borrowing from the Stoics, Pomponazzi thought that the realization that absolute security from natural disasters is elusive brought a greater mental security because of the recognition of an all-powerful God whose motives can never be fully understood. While Descartes was less explicit about God's role, he too pointed toward ethical purposes in *Les Météores*, declaring that understanding the true causes of the marvelous could eliminate stupefaction. Thus the intellectual changes that took place within natural philosophy during the sixteenth and seventeenth centuries were not the result of a simple replacement of Aristotelianism. Rather, as Descartes' *Les Météores* shows, the new natural philosophies, at times, paralleled the old, using their approaches, methods, explanations, and even goals while still seeking to distinguish themselves as novel and superior to the work of scholastics.

Notes

INTRODUCTION

1. Petrarca, *Invectives*, 64–69, 86–95.
2. Bausi, "Medicina e filosofia," 41–44.
3. Descartes, *Oeuvres*, 6: 231 (hereafter cited as AT). My translation.
4. "Ce qui me fait esperer que, si i'explique icy leur nature, en telle sorte qu'on n'ait plus occasion d'admirer rien de ce qui s'y voit ou qui en descent, on croya facilement qu'il est possible, en mesme façon, de trouver les causes de tout ce qu'il y a de plus admirable dessus la terre." Descartes, AT, 6:231. My translation.
5. Johnson, *Aristotle on Teleology*, 40–41.
6. Janković, *Reading the Skies*.
7. Bandini, *Clarissimorum Italorum epistolae*, 1:43. This letter is quoted in Giustiniani, *I tre rarissimi opusculi*, 12.
8. For variant readings of Spino's letter, see Fiorentino, *Studi e ritratti della rinascenza*, 141. For Porzio's psychology see Kessler, "Intellective Soul," 519–21.
9. Malagola, *Statuti delle università*, 274; Facciolati, *Fasti gymnasii patavini*, 325. For the distinctions among the various kinds of professorships, see Grendler, *Universities*, 144–46.
10. Frytsche, *Meteorum*, A6v.
11. Aristotle, *De generatione et corruptione*, 1.2.315b7–10, 317a13–18.
12. For a clear description of the warmth and coldness of air at various regions, see Albertus Magnus, *Meteora*, 1.1.7–1.1.10:10–15.
13. "Quid est Meteorologia? Est pars physices, quae considerat omnia corpora & species seu φαινομένα, quae fiunt & gignuntur in regionibus aeris vel terrae visceribus. . . . Quid est Meteoron? Meteoron est corpus imperfecte mixtum, ex vapore vel exhalatione generatum in aeris regionibus vi & calore radiorum coelestium." Frytsche, *Meteorum*, A6v. My translation.
14. For a variety of medieval and Renaissance positions about substantial forms, see Haas, "Mixture in Philoponus," 21–46.
15. "METEORON est aliquid, quod in sublimi aut in aëre accidit." Frytsche, *Meteorum*, 1. My translation.
16. "METEORA sunt res, seu ut Aristoteles appellat, πάθη, id est, affectiones, quae fiunt in sublimi, hoc est, in aere, & vicino elemento." Frytsche, *Meteorum*, 1. My translation.
17. Genuth, *Comets*.

18. Jenks, "Astrometeorology," 185–210; Hellmann, *Die Meteorologie.*

19. Theophrastus, *On Weather Signs,* §§ 19–21, p. 71.

20. Nifo, *De verissimis temporum signis.*

21. Guerlac, "Poets' Nitre," 243.

22. Heninger, *Handbook of Renaissance Meteorology.*

23. Linking current philosophical arguments and disciplines to Renaissance discussions is explicit in Hankins, introduction to *Renaissance Philosophy,* 6.

24. Garber, *Descartes Embodied,* 13–32.

25. Leijenhorst, *Mechanisation of Aristotelianism;* Newman, *Atoms and Alchemy;* Des Chene, *Physiologia;* Ariew, *Descartes;* Osler, "New Wine," 167–84.

26. Cook, "Cutting Edge?" 45–61; Park, "Natural Particulars," 347–67; Moran, *Distilling Knowledge;* Ogilvie, *Science of Describing,* 15–21.

27. Aristophanes, *Clouds,* 896.

28. For Thales' theory of earthquakes, see Seneca, *Naturales quaestiones,* 6.6.

29. Kusukawa, *Transformation of Natural Philosophy.*

30. Schmitt, *Aristotle and the Renaissance.*

31. Sorabji, "Ancient Commentators on Aristotle," 1–30.

32. For the tradition of noting Aristotle's obscurity during the Renaissance, see Schmitt, "Aristotle as a Cuttlefish," 60–72.

33. The essays in Sorabji, *Aristotle Transformed,* illustrate the philosophical originality of the ancient commentary tradition. The intellectual vitality of the Renaissance commentary tradition is maintained in the follow works: Nardi, *Saggi sull'Aristotelismo;* Randall, *School of Padua;* Kristeller, *La tradizione aristotelica;* Poppi, *Introduzione;* Schmitt, *Studies on Renaissance Aristotelianism;* Schmitt, *Aristotle and the Renaissance;* Schmitt, "History of Renaissance Philosophy," 9–16; Lohr, *Latin Aristotle Commentaries;* Mercer, "Early Modern Aristotelianism," 33–67; Hellyer, *Catholic Physics.*

34. I use the term *chymical* to refer to early modern chemistry and alchemy in order not to make a sharp anachronistic differentiation between the two fields. See Newman and Principe, "Alchemy vs. Chemistry," 32–65.

35. Bacci, *Ne' quali si tratta della natura,* 259–60; Trevisi, *Sopra la inondatione del fiume.*

CHAPTER 1: THE EPISTEMOLOGY OF METEOROLOGY

1. Lines, "Natural Philosophy," 267–320.

2. Lohr, *Latin Aristotle Commentaries,* 282–86, 331–32, 347–62.

3. "Aristotle argued for a rational search for certain causes in medicine as everywhere else." Franklin, *Science of Conjecture,* 166.

4. "The lure of demonstrative certainty drew scholastic natural philosophers to believe that they could make knowledge that was analytically solid." Dear, *Revolutionizing the Sciences,* 5–6. "According to most Christian Aristotelian philosophers including Thomas Aquinas, one should demand absolute certainty of any religious or natural scientific statement that commands assent." Olson, *Science and Religion,* 90. See also Cook, *Matters of Exchange,* 15–16.

5. "The shift from an Aristotelian theory of knowledge, which was confident of the truth of what we perceive, to a modern skepticism that doubts our capacity to know truth but nonetheless finds the resources, intellectual and cultural to overcome pessimism and to insist that scientific knowledge is still attainable." Jacob, *Scientific Revolution*, xiii.

6. Shapin and Schaffer, *Leviathan*, 23–24.

7. Shapiro, *Probability and Certainty*, 3–14.

8. Cassirer, *Das Erkenntnisproblem*, 136–44; Randall, "Development of Scientific Method," 177–206; Randall, *School of Padua*.

9. Palmieri, "Giacomo Zabarella," 418.

10. Nifo, *De physico auditu*, 6v.

11. Shapin and Schaffer, *Leviathan*, 24.

12. Aristotle, *Posterior Analytics*, 1.7.75a38–b6.

13. Aristotle, *Nicomachean Ethics*, 1.3.1094b13–27.

14. Moss and Wallace, *Rhetoric and Dialectic*, 16.

15. Serjeantson, "Proof and Persuasion," 139–40.

16. Freeland, "Scientific Explanation," 62–107.

17. Serjeantson, "Proof and Persuasion," 139.

18. Aristotle, *Metaphysics*, 6.2.1027a12. Translation from Aristotle, *Complete Works*, 2:1621–22.

19. Aristotle, *Posterior Analytics*, 1.6.75a19–22.

20. For similar arguments as applied in Renaissance medicine, see Maclean, *Logic, Signs and Nature*, 132.

21. Aristotle, *Meteorology*, 1.1.338b1–3.

22. Aristotle, *Meteorology*, 1.1.339a3–4.

23. Aristotle, *Meteorology*, 1.7.344a5–7.

24. Morrison, "Philoponus and Simplicius," 1–22, see especially p. 6; Viano, *La matière*, 189–90.

25. Freeland, "Scientific Explanation," 62–107.

26. Viano, *La matière*, 190.

27. Aristotle, *Meteorology*, 1.4.342a31–34.

28. Aristotle, *Meteorology*, 2.3.359a11–15.

29. Aristotle, *Meteorology*, 2.8.366a1–2.

30. Aristotle, *Meteorology*, 2.8.366b20–30.

31. Aristotle, *Meteorology*, 2.8.366b31–367a11, 2.8.367a21–b8.

32. Hacking, *Emergence of Probability*, 39–45.

33. Albertus Magnus, *Meteora*, 1.1.1:3.

34. Buridan, *Expositio libri meteororum*, fol. 103r.

35. "Forma, quae Substantia sit, proprie Metheoris non competit, quia sunt mixta imperfecta, quae huius sunt conditionis, ut nova non prodierit, ideo forma Substantialis non est distincta ab ea Elementorum." Piccolomini, *Librorum ad scientiam pars quarta*, 4v. My translation.

36. Schegk, *In reliquos naturalium*, 335; Vimercati, *In quatuor libros meteorologicorum*, 6r; Conimbricense, *In meteorologicos*, 5.

37. Plato, *Timaeus*, 47e–48e.

38. Plato, *Timaeus*, 29b–c.

39. "Materia est causa, quod inveniatur in scientiis id, quod est per accidens: et quod est per accidens est magna elongatio a veritate . . . oportet igitur ut materia sit minus vera." Averroes, *In librum de demonstratione*, 1.2:375r. My translation.

40. "Incertitudo causatur propter transmutabilitatem materiae sensibilis; inde quanto magis acceditur ad eam, tanto scientia est minus certa." Thomas Aquinas, *In libros posteriorum*, Lib. I, lect. 41, no. 3, p. 58. My translation.

41. "In quibus non omnia perfecte et secundum certitudinem tradere possumus, sed quaedam sub dubitatione relinquemus." Thomas Aquinas, *In libros meteorologicorum*, Lib. I, lect. 1, no. 7, p. 393. My translation.

42. Olivieri, *Certezza e gerarchi del sapere*, 67–69.

43. Porzio, *De humana mente, disputatio.*

44. "Verum respectu aliquorum, ut puta quarundam elementorum affectionum, quas in libro Meteorum alias enarravimus, est causa particularis, hoc est, est proxima & immediata efficiens causa: . . . Atque iccirco in rebus naturalibus demonstrationes a causis efficientibus ad suos effectus non sunt potissimae: quia effectus impediri queant, ob materiae indeterminationem, quae varias multiplicesque formas recipere potest. Porro aut necessitas quae a materia oritur Philosophis appellatur necessitas simpliciter: nullam aliam ob rationem, nisi quod esse rerum naturalium a materia oriatur, Neque aliter asserere potuerunt quod non agnoverint aliam substantiam, praeter materiam." Porzio, *De rerum naturalium principiis*, Riiv–Riiir. My translation.

45. "Quae essent quaedam ingenerabilia prorsus & incorruptibilia certas leges sibi descriptas inviolabiliter observantia, quas non praetergrediantur in secula seculorum ex quibus sunt nimirum coelestes spiritus, atque virtutes, sol, luna, stellae, atque coelorum orbes. Sub quibus prospexit quaedam nullam habentia determinatam legem suae generationis, sed verbo Dei stare tantum: ut dum ille iusserit, & quando iusserit gignantur & corrumpantur. Si enim dixerit altissimus, statim ignis de coelo, aut pluvia, aut grando, aut ros, aut pruina, aut fulgur, aut tonitru descendet: & cum placuerit illi, statim cessabunt." Titelmans, *Compendium*, 146–47. My translation.

46. "Haec omnia tractavimus, & explicasse nonnulla videbimur: tum enim Deus revelasti nobis opera tua. / Veruntamen quae adhuc occulta sunt nobis, plurima sunt: & parva nimis est portio eorum quae novimus, si conferatur ad ea quae servantur ocultata a nobis. / Quae cognoscimus sunt pauca nimium, & ea ipsa quae forte cognovimus, sunt heu nimium tenuiter a nobis cognita. / Nimium tenuis est & obscura scientia nostra, neque est vel unum ex cunctis operibus altissimi, quod sit pro dignitate sua nobis cognitum." Titelmans, *Compendium*, 163.

47. Avempace, *Aristotle's Meteorology*, 399.

48. "Accidit autem huic cognitioni in substantia Galasiae esse cognitionem diminutam: propterea quod genus illius est ignotum esse per se." Averroes, *In quatuor meteorologicorum libros*, 5:414r.

49. Pomponazzi, *De incantationibus*, 6. On saving the appearances in Pomponazzi, see Maclean, "Heterodoxy in Natural Philosophy," 15–16; Graiff, "I prodigi e l'astrologia," 331–32. My translation.

50. Pomponazzi, *De incantationibus*, 130–31. My translation. For raining wool see Pliny, *Historia naturalis*, 2.57.

51. Pomponazzi, *De incantationibus*, 81. My translation.

52. Aristotle, *Meteorology*, 2.1.353b25−29.

53. "Et hoc est contra experientiam; nam, ut multi ex scolaribus mihi dixerunt vidisse, et ego ipse vidi multos fontes et puteos stantes naturales naturaliter absque aliquo artis ministerio." Pomponazzi, *In libros meteororum*, 49r. My translation.

54. "Sed forsan in patria Arist. ita erat quod omnes aquae stationariae erant manufactae. In nostris tamen regionibus non ita est." Pomponazzi, *In libros meteororum*, 50r. My translation.

55. "Hoc autem est probabile, et non demonstrat." Pomponazzi, *In libros meteororum*, 49v. My translation.

56. "Nam meo iudicio ratio Aristo. nullius est valoris." Pomponazzi, *In libros meteororum*, 50r. My translation.

57. "Nollem ego videri temerarius. nam haec non dico ut velim dicta Arist. reprehendere. sed tamen circa illa absque reprehensione mihi concessum et licitum est dubitare; nam meo iudicio ratio Aristo. nullius est valoris. Ad hanc rationem scio ego respondere nec eam novi, et [ut written above] respondere velim. Forsan dicendum est, quod Ar. non protulit hanc rationem pro demonstratione et necessaria probatione, sed ut esset quaedam persuasio verisimile. Enim videtur quod non possit mare a fontibus ortum trahere ratione illa, quam ponit Ar. non autem demonstrat. id dicerem Aristotelem non potuisse pro demonstratione. Valet ergo a verisimili, non absolute proferendo. absolute enim proferendo falsa esset. sed Ar. assumpsit eam ut in pluribus non absolute." Pomponazzi, *In libros meteororum*, 50r. My translation.

58. "Alia dubitatio satis trivialis: quomodo scivit Aristoteles tiphonem illo modo contrarietate, cum non fuerit supra nubem, nec viderit eum generari. Notetis quod tantum petendum est ab auditore, quantum concedit materia tractata: et clare habetur primo Aethic. [i.e., *Nicomachean Ethics*] et in meth. [i.e., *Metaphysics*]. In rebus enim naturalibus non possumus habere demonstrationem semper sic ut diximus in lib. de Anima, meth., autem prius, quoniam nemo possit scire illa diurna superiora. Nos enim sumus ut manuales, Deus vero ut Architectus, possumus enim nos facere artificialia non autem naturalia, unde in hac scientia possumus habere demonstrationem, et certitudinem eo modo quo possumus, non autem ut in meth., Sed quia Aristoteles remotior est a contradictione, ideo sua dicta in naturalibus probantur." Pomponazzi, *In libros meteororum*, 84r. My translation. For Pomponazzi's claims of using an uncertain method in other works, see Perfetti, "Docebo vos dubitare," 439−66; Raimondi, "Pomponazzi's Criticism of Swineshead," 326.

59. "Explicare quot e mixtionibus contingant significationes laboriosum est: et fortasse supra captum humanum: quapropter observatoribus pensitanda relinquimus." Nifo, *De nostrarum calamitatum causis*, 18r. My translation.

60. For medicine, see Maclean, *Logic, Signs and Nature*, 134.

61. "Tum etiam quia cause elementarie e quibus, hec proficiscuntur, contingenter agunt, et contingenter patiuntur." Nifo, *In libros meteorologicorum*, 8.

62. "Quoniam autem de effectibus sensui immanifestis putamus sufficienter demonstrasse secundum rationem si ostensa de his reduxerimus ad possibilem, hoc est ad talem certitudinem, ad quam non sequantur impossibilia." Nifo, *In libros meteorologicorum*, 116.

63. "Licet aliqui naturales effectus sint manifestiores, et aliquid minus manifestiores, omnes sunt certissimi, saltem quo ad quia est." Nifo, *In libros meteorologicorum*, 117.

64. "Dicendum, scientiam de natura non esse scientiam simpliciter, qualis est scientia mathematica, est tamen scientia propter quid: quia inventio causae, quae habetur per syllogismum coniecturalem, est propter quid effectus. per haec delentur obiectiones, quae contra haec fieri solent: Prima quidem delentur ex eo, quia non est circulus in demonstratione, cum primus processus sit tantum syllogismus, secundus vero demonstratio propter quid. deletur etiam Secunda obiectio, quia effectus semper est notior ipsa causa in genere notitiae quia est. nunquam enim causa potest esse ita certa quia est, sicut effectus, cuius esse est ad sensum notum. Ipsum vero quia est causae, est coniecturale, utrum tale esse coniecturale est notius ipso effectu, in genere notitiae propter quid. nam posita inventione causae semper scitur propter quid effectus. unde & Aristo[teles] in libro Meteororum concedit se non tradidisse veras causas effectuum naturalium, sed quo erat sibi possibile coniecturabiliter." Nifo, *De physico auditu*, 6v.

65. "Suppositiones autem ex se patent; prima quidem ex communi consensu et experientia, cum in locis montuosis signa adhuc inundationis appareant atque vestigia, ut conchae et ostrea, ut deveniamus iure in coniecturam quandoque diluvium fuisse universale, montuosa quaecumque cooperiens ac tegens." Russiliano, *Apologeticus adversus cucullatos*, 154.

66. "Quaesita non pauca de rebus naturalibus existunt, quae variis solvuntur rationibus, quarum multitudo provenit ob ignorantiam verae & propriae causae. multis etiam non habetur fides, aut certe minus eis credimus, aut non tanquam necessariis assentimur, quia videlicet sensum effugiunt, & ob id pro minus necessariis habentur." Schegk, *In reliquos naturalium*, 333.

67. "Liquet hactenus, ea quae fiant & generentur, ut sunt res omnes materiatae, indagatorem naturae non considerare quando et quomodo fiant necessario (illud enim in his rebus sciri nequit) sed quando & quomodo possint fieri. nam talis cognitio est, quales ipsae quarum ortus obitusque contigentes esse experimur." Schegk, *In reliquos naturalium*, 333.

68. Duhem, *Sozein ta phainomena*.

69. Barker and Goldstein, "Realism and Instrumentalism," 232–58.

70. "Et lex observanda est ista, quod cum causae alicuius effectus sunt nobis ignotae, debemus accipere suppositiones, aut principia, ex quibus nihil impossibile sequatur, neque contra sensum, neque apparentibus repugnans. . . . Propositio possibilis contingens est illa quae non est vera, sed potest esse vera, ut habetur primo priorum. & Averr. hoc contingens appellat inventum, & contingens & possibile idem sunt apud Graecos." Boccadiferro, *Super primum librum meteorologicorum*, 45v–46r.

71. Martin, "Experience," 1–15.

72. Aristotle, *Meteorology*, 2.1.354a20. "Observatus vero est & alius in mari fluxus, tum in Mediterraneo, tum in Oceano, quo videlicet fluit ab ortu ad occasum, & in Mediterraneo rursus ad ortum refluit, quomodo etiam in sinu illius Adriatico. Quem fluxum, etsi non evidentem, observaverunt tamen nautae ex itineribus, quae breviori tempore conficiunt, cum ab ortu ad occasum navigant, quam cum ab occasu ad ortum, aquae fluxu navium motum aut adiuvante, aut impediente." Vimercati, *In quatuor libros meteorologicorum*, 74v.

73. Zabarella, *De rebus naturalibus*, 386–87. See also Schmitt, "Experience and Experiment," 98–100.

74. Martin, "Aristotelians," 135–61.

75. "Incipit, ut dixi, probare a posteriori, seu potius Methodo resolutoria, ostendere, bene assigatam esse causam terraemotus." Cabeo, *Commentaria*, 2:246. My translation.

76. Cabeo, *Commentaria*, 2:243. The idea that the causes of earthquakes were chymical did not originate with Cabeo; see Agricola, *De ortu & causis subterraneorum*, 31–32.

77. Cabeo, *Commentaria*, 2:243.

78. Cabeo, *Commentaria*, 2:248.

79. Cabeo, *Commentaria*, 2:249. The sulfuric, bituminous, and nitric nature of the area around Vesuvius was widely noted by observers of its 1631 eruption and previous eruptions. E.g., Spinola, *Discorso*, 4; Porzio, *De conflagratione*, 6.

80. Frytsche, *Meteorum*; Serjeantson, "Proof and Persuasion," 160.

81. "Numerus elementorum, quamvis varius fuerit apud Antiquos, ut dicitur in Phys. communiter tamen censetur esse quaternarius; verum est tamen, quod non datur ratio aliqua evidenter demonstrans elementa non esse plura, nec pauciora, & praesertim rationes Arist. sunt probabiles." Mastri and Belluti, *Disputationes*, 203.

82. "Haec ratio iam nihil concludit, sed nos remittit ad discutiendas totius Physicae difficultates, ut tandem constet an eas dirimat accuratius Aristoteles, an antiqui: & vere opinio cuiuscunque philosophi nihil est aliud quam hypothesis, qua posita videndum est an facilius enodentur omnes difficultates scientiae naturalis, ut fieri solet in Astrologia quae varios statuit epicyclos, concentricos & eccentricos, ut iis quae apparent in coelo respondeatur: sic arbitrandum est de opinione Aristotelis & antiquorum, nec ante sententiam ferre oportet, quam praecipuis dubitationibus responsum fuerit." Bérigard, *Circulus pisanus*, 19. My translation. Cf. Marangon, "Aristotelismo e cartesianesimo," 4.2:95–114.

83. Waddell, "World," 3–22.

CHAPTER 2: TELEOLOGY IN RENAISSANCE METEOROLOGY

1. Bacon, *De augmentis scientiis*, 1:571; Descartes, *AT*, 1:15; Spinoza, *Collected Works*, 439. For the importance of the rejection of final causes in the historiography of the scientific revolution, see Koyrè, *Newtonian Studies*, 7–8; Shapin, *Scientific Revolution*, 28–30. For an overview of the scientific revolution and scholastic views of teleology, see Johnson, *Aristotle on Teleology*, 23–30.

2. Osler, "From Immanent Natures," 388–407; Osler, "Whose Ends?" 151–68; Shanahan, "Teleological Reasoning," 177–92; Simmons, "Sensible Ends," 49–75.

3. Diderot and d'Alembert, *Encyclopédie*, 2:789.

4. Des Chene, *Physiologia*, 168–211.

5. For the view that seasonal rains have final causes, see Furley, "Rainfall Example," 115–20. For the position that these purposes are anthropocentric, see Sedley, "Is Aristotle's Teleology Anthropocentric?" 179–96. For the rejection of their view, see Wardy, "Aristotelian Rainfall," 18–30; Johnson, *Aristotle on Teleology*, 149–58.

6. Kusukawa, *Transformation of Natural Philosophy*.

7. Taurellus, *Alpes caesae*, 99. For Taurellus's subsequent influence on this issue, see Johannes Schreiner's dissertation from the University of Leipzig, *Disputata meteorologica*, B4r.

8. Petersen, *Geschichte der aristotelischen Philosophie*, 219–58.

9. Aristotle, *De generatione et corruptione*, 2.10.336a33–b6.

10. Aristotle, *De generatione et corruptione*, 2.10.336b22.

11. Aristotle, *De generatione et corruptione*, 2.10.336b31–32.

12. Aristotle, *De generatione animalium*, 5.1.778a32–b2.

13. For possible discussions of final causes, see Aristotle, *Meteorology*, 2.3.359a15, 2.3.359b15; Viano, *La matière*, 143.

14. On *Meteorology* 1–3 as about inanimate homeomerous substances, see Olympiodorus, *In Aristotelis meteora commentaria*, 20–21, 273.

15. For the need for intelligent agents for final causes, see Des Chene, *Physiologia*, 186–200; Menn, "Dennis Des Chene's *Physiologia*," 122. For examples of Renaissance Aristotelians referring to the realization of form as a final cause, see Nifo, *In libros meteorologicorum*, 130: "Finis generationis comete est substantialis forma"; Boccadiferro, *In secundum, ac tertium meteororum*, 33r: "Finis generationis est forma rei generatae."

16. See Aristotle, *Physics*, 2.8.198b16–21. Whether Aristotle actually endorsed a teleological position in this case has been a matter for debate. See Furley, "Rainfall Example," 115–20; Sedley, "Is Aristotle's Teleology Anthropocentric?" 179–96; Wardy, "Aristotelian Rainfall," 18–30.

17. For the prominence of natural philosophy that was based on Aristotelian texts in Renaissance Italian universities, see Grendler, *Universities*, 267–313. For the importance of paraphrastic writing on Aristotle in France, see Kessler, "Lefèvre Enterprise," 1–22; Rice, "Humanist Aristotelianism," 132–49.

18. For the separation of philosophy from, as well as its subordination to, theology in the Italian tradition, see Maclean, "Heterodoxy in Natural Philosophy," 1–29.

19. Averroes, *In quatuor meteorologicorum libros*, 5:403r–4v.

20. "Tractatus primus de causis impressionum omnium, materiali et efficiente." Albertus Magnus, *Meteora*, 1.1.1:1.

21. Gaetano of Thiene, *In quattuor Aristotelis metheororum libros*.

22. "Quod ideo facit, ut eiusmodi rerum, quae in sublimi oriuntur, principia, & causas communes, quae duae sunt, materia, & agens, scrutentur." Vimercati, *In quatuor libros meteorologicorum*, 6v.

23. Cesalpino, *Peripateticarum quaestionum libri quinque*, 21v, 64v–67r.

24. Ramberti, "Stoicismo e tradizione," 51–84. For an excellent account of the historiography of Pomponazzi, see 51–62. For the foundations of modern scholarship on Pomponazzi, see Nardi, *Studi su Pietro Pomponazzi*.

25. For Pomponazzi's debts to medievals, see Schabel, "Divine Foreknowledge," 165–90. For Pomponazzi's alleged role in initiating modern science, see most recently Gaukroger, *Emergence*, 107–16.

26. "Dicit enim philosophus quod homogenea ista, ut metallica, sunt minus nota quam organica: non quia sint in composito, vel extra compositum; sed quia minus habent de forma & plus de materia: viciniora enim sunt materiae magis quam distantia a forma." Pomponazzi, *Dubitationes*, 50r. My translation.

27. "Nulla enim res abstracta est in mundo quae non conveniat naturae ad aliquid, et propriam habeat utilitatem in universo, et in suo genere sit maxima bona: Deus enim secundum Philosophos est auctor optimus et sapientissimus, cum autem universum sit opus Dei, oportet ergo quod perfectissime hoc fecerit, ut Plato posuit in Thimeo." Pomponazzi, *In libros meteororum*, 77v. My translation.

28. Seneca, *Naturales quaestiones*, 6.18,1.

29. "Unde Divus Dionisius in quarto cap. 4 de divinis nominibus, illi canes non sunt boni, qui non sunt furiosi, sic venti fortissimi sunt optimi in suo genere, et asinus qui est grossissimae naturae inter omnes asinos est optimus asinus. Unde venti Turbines non essent turbines nisi tales essent sicut sunt, unde relucet maximi Dei perfectio in illis, sicut etiam fulmen non esset fulmen, nisi operaretur ea, quae operatur: sic etiam terrae motus non esset terrae motus, nec bonus in suo genere nisi subverteret civitates et provincias." Pomponazzi, *In libros meteororum*, 79r–79v. See also Pseudo-Dionysius Areopagita, *De divinis nominibus*, 4, 25.

30. "Deus enim est causa omnium rerum, excepta mala voluntate, huius enim nos causa sumus, multa enim nobis videntur mala, quae sunt optima, quoniam nos finem illorum ignoramus." Pomponazzi, *In libros meteororum*, 80r. My translation.

31. "Unde forsan esset bonum, quod Turca veniret, quoniam essemus, postea meliores christiani." Pomponazzi, *In libros meteororum*, 80r–80v. My translation.

32. "Ideo Philosophi sunt securissimi, quoniam ipsi sciunt haec omnia ex ordinatione naturae evenire, ideo non mirantur de illis effectibus, ut facit ignobilis vulgus; quoniam horum effectuum causas cognoscunt; et ita esse ordinatum optime a natura, Ideo sciunt se positam et ordinationem Dei." Pomponazzi, *In libros meteororum*, 80v. My translation.

33. Pomponazzi, *De incantationibus*.

34. "Profecto, Domini mei, non est mirum si Philosophi irridentur a vulgaribus, quoniam volunt omnia scrutari, et quid possit Deus, quoniam non grosso modo intendunt in via populari, sed volunt secreta Dei uniari, et naturae, Ideo spernunt divitias et voluptates: exemplo fuit concivis noster Petrus de Mantua, qui mortuus est in Hospitali, unde altiora re ne quaesiveris. Ideo bene et melius faciunt isti Religiosi, qui securius respondent, quoniam voluntas Dei sic vult, ideo sic fit, Ideo non est alia illorum quaerenda causa. Peripathetici autem et alii stulti Philosophi qui volunt omnia scire, dicunt ex necessitate motus Coeli haec evenire, quoniam necesse est omnia quae sunt genita aliquando interire, et alia rursum gigni." Pomponazzi, *In libros meteororum*, 54r–54v. My translation.

35. "Dico quod hoc fecit propter melius scilicet bonum universi, quod postea est maior perfectio . . . aut quia sic placuit, ut bene dicunt Theologi nostri, Omnia enim quae voluit Deus fecit in Coelo, et in terra." Pomponazzi, *In libros meteororum*, 57r. My translation.

36. "Philosophi tamen possunt et Astrologi hoc salvare alio modo licet dicta sua non sint vera, nam hoc potest esse ex corporibus coelistibus, et ex causa motus syderum." Pomponazzi, *In libros meteororum*, 81r. My translation. Cf. Graiff, "I prodigi e l'astrologia," 331–61.

37. "Primo modo tunc sunt per se intenta, secundo modo non sunt per se intenta: nam quod sint duo oculi duo nares haec omnia habent suum finem, sed quod homo nascatur claudus, aut quod sint pili sub ascellis, illa magis videntur nocere, quam prodesse, ideo dicuntur esse ex necessitate materiae." Pomponazzi, *In libros meteororum*, 107v.

38. "Et haec quaestio satis difficilis est, nam primo quod fiant ex necessitate materiae, et non propter finem, quoniam ex secundo de generatione et octo physicorum se motu solis et astrorum generationes et corruptiones fiunt in istis inferioribus, nam isti effectus refractorum sunt coniuncti aliis affectibus, nam quia sol sic movetur, et invenit talem materiam sic dispositam ideo necessario agit, ideo non videtur, quod hoc sit factum ex intentione, sed sicut quando pluit sub cane, quod casuale est, sic videtur etiam dicendum de istis, fit enim in pauciori, nam rarissimi fiunt illae refractiones, cum raro videamus iridem, rarius autem parelias et virgas." Pomponazzi, *In libros meteororum*, 107v. My translation.

39. "Ad aliam partem, quod sint per se intenta a natura, quoniam habemus in fine tertii de anima, in libro de animalibus in libro de partibus et mille aliis in locis, quod natura nihil facit frustra. Sed illis sunt effectus naturales ergo non fiunt frustra. Item primo posteriorum [*Analytics*] et primo meth.[i.e., *Metaphysics*] ea quae sunt per se sunt determinata, et debent fieri semper aut frequenter, sed illi effectus fiunt semper aut frequenter, casus autem fiunt extra semper, aut frequenter, ergo ista non sunt casualia, quod autem illi sint effectus per se, patet nam Aristoteles vult se probasse effectus istos demonstratione, et mathematice, ideo certissime ergo amplius in primo posteriorum ubi dicit eorum quae saepe fiunt sunt demonstrabiles." Pomponazzi, *In libros meteororum*, 108r. My translation.

40. "Ibi omnes expositores dicunt quod sed illa raro fiant in ordine ad tempus. Tamen in ordine ad suam causam saepe, imo semper fiunt, nam numquam causa est disposita quin semper fiat aut sequatur effectus, nam numquam sit interpositio terrae inter solem et lunam, quia sequatur eclipsis." Pomponazzi, *In libros meteororum*, 108r. My translation.

41. For this debate, see Hasse, "Spontaneous Generation," 150–75.

42. "Excrementa scilicet per se non sint intenta a natura, tamen natura utitur eis ad aliquod bonum, nam nos reiicimus stercora, et rustici colligunt ea." Pomponazzi, *In libros meteororum*, 108v. My translation.

43. The view that Pomponazzi attributes to Averroes may not be the one that he held. Averroes held that everything possesses a final cause. See Belo, *Chance and Determinism*, 138–39.

44. Aristotle, *Physics*, 2.4.195b36–196a5.

45. "Et ideo dicit Averroes quod falsum est, quod scribitur in lege Maumethi, quod omnia quae habent evenire necessario eveniant; nam nec Deus, nec alius propter Deum hoc cognovit." Pomponazzi, *In libros meteororum*, 109r.

46. "Aliae enim omnia quae accidunt de necessitate evenirent. Non eveniunt autem illa de necessitate, sed a casu, et ideo dicit Averroes quod falsum est, quod scribitur in lege Maumethi, quod omnia quae habent evenire necessario eveniant; nam nec Deus, nec alius propter Deum hoc cognovit. Sed notetis quod D. Thomas et alii dicunt, quod illa coniunctio non habet causam, tamen Deus hoc ponit, quoniam videt omnia: et hoc quod dicit S. Thomas est verissimum, quoniam nihil est simpliciter casuale, quoniam omnia sunt Deo nota, et causas habet Deus notas et determinatas, sed naturaliter hoc est casuale." Pomponazzi, *In libros meteororum*, 109r. My translation. For Thomas's rejection of chance as a cause, see *Summa theologiae*, I, q. 103, a.1.

47. For Pomponazzi's pessimism toward the possibility of complete knowledge of the natural world, see Perfetti, "Docebo vos dubitare," 439–66.

48. "Haec autem dicta sunt Theologia et bene dicta, sed non sunt ad mentem Aristotelis. Multa autem hic magis credenda sunt, quam investiganda, sed stulti Philosophi velint omnia investigare. Aristoteles enim vult quod illa coniunctio nullam habeat causam, et ideo illud quod dicunt isti viri non est Peripateticum nec Accademicum." Pomponazzi, *In libros meteororum*, 109r. My translation.

49. "Quae denique talium rerum config potest perfectio, cum aut mala portendunt, aut magna hominum calamitate, illa secum afferunt? iam vero, quae tandem est ista perfectio? Cum non modo quicquam ad universi aut ordinem, aut ornatum, aut utilitatem (quae tres dotes mundum praecipue muniunt) conferat, sed potius omnia perturbet, decore exuat, atque demoliatur." Galesi, *De terraemotu*, 69. My translation.

50. On the details of Galesi's academic career, see Lines, "Natural Philosophy," 316.

51. "Dico ulterius, quod licet multae sint impressiones quarum ignoramus finem: tamen bene de eis potest haberi scientia, cum habeant veras, firmas determinatas causas: quibus positis, de necessitate sequitur effectus ille." Boccadiferro, *Super primum librum meteorologicorum*, 2r.

52. "Ultima dubitatio de causa finali quam omittit Aristo. quid ergo finis cur non expressit super, dico ex libris physicorum, quia duplex finis generationis, & rei generatae, finis generationis est forma rei generatae, finis rei generatae perfectio universi, finis ergo generationis est exhalationis finis rei generatae est perfectio universi, finis vero specialis multiplex est, quia est navigatio." Boccadiferro, *In secundum, ac tertium meteororum*, 33r.

53. "De causa finali dico quod aut terremotus non habet causam finalem, fit ex necessitate materiae iste effectus, non est intentus exhalatio non intendit istum motum, sed secundum ascendentem, aut si habet finem, finis generationis motus terrae, quia permet fumidae ehxalationis [*sic*] acquiritur iste motus finis rei generatae, est perfectio universi." Boccadiferro, *In secundum, ac tertium meteororum*, 38v. My translation.

54. Seneca, *Naturales quaestiones*, 2.32.

55. Aristotle, *Meteorology*, 1.7.344b19–345a5.

56. Aristotle, *Meteorology*, 1.6.343b2–3.

57. Genuth, *Comets*, 17–50.

58. Smoller, "Of Earthquakes," 156–87.

59. For example, see Ginzburg, *Night Battles*, 23.

60. Barnes, *Prophecy and Gnosis*, 58.

61. Scheel, *Dokumente zu Luthers Entwicklung*, 151.

62. Luther, *Luthers Werke kritische*, 11:207; 29:621–22; 32:228–29; 37:616; 45:338. I thank Andrew Sparling for bringing these passages to my attention.

63. Luther, *Luthers Werke kritische*, 3:364–65; 4:365–66.

64. Leppin, *Antichrist und Jüngster Tag*, 87–96.

65. For Melanchthon and the development of Aristotelian "textbooks" in the sixteenth century, see Schmitt, "Philosophical Textbook," 792–804; Reif, "Textbook Tradition," 17–32.

66. "Quare teneamus quod verissimum est, esse multarum rerum in natura veram & certam cognitionem, etiamsi non omnia pervestigari possunt. Haec cum supra dicta sint de certitudine, nunc sum brevior, etsi tantum ideo de modo procedendi disseritur, ut certitudo quaeratur." Melanchthon, *Initia doctrinae physicae*, 22r.

67. Methuen, *Kepler's Tübingen*, 90–91.

68. Kusukawa, *Transformation of Natural Philosophy*, 132.

69. Pontano, *I poemi astrologici*.

70. Nauert, "Caius Plinius Secundus," 384.

71. "Nihil autem opus est me de Plinii laudibus dicere, qui non tam librum, quam bibliothecam, nobis reliquit, complexus uno volumine fere totam rerum naturam: ut ne apud Graecos quidem extet unus aliquis locupletior autor. Adhaec, conservavit multis de rebus lectissimas doctorum sententias, item Latinas appellationes plantarum." Milich, *De mundi historia*, 9r.

72. "Quare elegimus hunc librum Plinii, qui quasi in compendium contraxit praecipuos locos universae Physices: inclusit enim breviter in hunc unum libellum multa Aristotelis Volumina, librum de coelo, μετέωρα, de mundo, addit & plaeraque Astronomia, necessaria in tali opere, quae tamen non attingit Aristoteleles, credo quod Graecis nondum satis noti essent planetarum motus." Milich, *De mundi historia*, 7r.

73. "Hanc Epicuri sententiam secutus hoc loco Plinius coniungit atque commiscet impressiones in aere factas, cum apparentiis coelestibus, sed ita tamen, ut prudens lector facile vera a falsis discernere possit praesertim si mediocriter in philosophia Peripatetica versatus fuerit, quae in his naturalibus disputationibus longe vincit reliquas." Milich, *De mundi historia*, 84v.

74. "Quia vero universa in hac natura destinata sunt ad certos fines, ergo necesse est ut hi fines pendeant ab aliqua mente gubernante hanc totam naturam." Milich, *De mundi historia*, 86r. My translation. For the connection between the denial of providence and Epicureanism in Lutheran theology, see Preus, *Post-Reformation Lutheranism*, 2:196.

75. "Physicus autem quia tantum caussas in materia quaerit, itaque et fines tantum circa materiam quaerit, videlicet omnes istas impressiones significare aliquam aeris mutationem vel serenitatem vel tempestatem etc." Milich, *De mundi historia*, 86r. My translation.

76. "Hactenus de physicis effectibus Cometarum dixi, de tempestatibus & aeris viciis. Haec non significat tantum sed efficit." Milich, *De mundi historia*, 93v. My translation.

77. "Cum igitur omnium aetatum observatione compertum sit, Cometas minari tristes eventus, recte in caussa finali haec significatio recensetur." Milich, *De mundi historia*, 95r. My translation.

78. "Prodigiosa pluvia dicitur, quando cum pluvia aliquando decidunt vermes, ranae, pisces, lac, pili. Item cum pluisse lapidibus, carne, sanguine, ferro & aliis rebus apud historicos legitur. Quarum rerum potest quidem naturalis ratio assignari, sed non adeo inexpugnabilis. Nam eas esse fatales & miraculosas pluvias, & quasi signa credibile est, & inter portenta referri possunt." Frytsche, *Meteorum*, 25v. My translation.

79. "Est enim pluvia ut flumen divinae providentiae, qua providet Deus vegetabilibus, ut moderate humectentur, & quasi clepsidra irrigentur, aqua guttatim e summo labente." Frytsche, *Meteorum*, 30r.

80. "Finis Comentarum est parare siccitatem, pestem, famem, bella mutationem regnorum & Rerumpub. legum, traditionum." Frytsche, *Meteorum*, 93r. My translation.

81. Frytsche, *Meteorum*, 120r, 147r.

82. "Possuntne omnium Meteorum monstrari causae Physicae? Non. Affirmat enim sacra scriptura, multa oriri a prima causa rectrice & moderatrice universae naturae incompraehensibili & infinita, praeter consuetum & solitum modum, quae liberrima voluntate sine connexione stoica secundarum causarum, multa perficit, movet exacuit, reprimit, compescit, mutat & excitat, quae insolitorum & singularium casuum sunt praenuncia." Garcaeus, *Meteorologia*, 7r. My translation.

83. "Primo enim haec doctrina Stoicos refutat, qui contingentiam & providentiam tollunt, & docet etiam contra Epicuros, hunc mundum non regi casu, nec omnia fieri necessario, nec fortuito exoriri Meteora, quorum alia producuntur ideo, ut sint signa futurorum eventuum, qualia sunt parelia, irides, halones: partim ut sint quasi instrumenta, quibus hominum ingravescentem malitiam puniat DEUS, ut sunt grandines, procellae, fulmina, terraemotus, Cometae &c. partim etiam ut terram suaviter rigent, aut nimium humecatam exiccent, ut sunt pluviae, nives, ros ventorum flatus & similia Meteora." Garcaeus, *Meteorologia*, 12v.

84. "Est autem duplex causa finalis Meteororum. Alia est Physica, alia Theologica. Primo enim Physice loquendo, aeris mutationem significant, serenitatem, tempestatem, humiditatem vel siccitatem. In igneis etiam salus animalium spectantur. Nam per haec fumi sublati consumuntur, ne putredine omnia suffocentur & extinguantur, qualis & eventationis usus

est in corpore humano. Quin & aqueis homines carere non possunt, pluvia, fluminibus fontibus, nubibus, ventis, non tantum ut aqua effundatur & colligatur, sed etiam ut Solis ardor aliquando mitigetur, terra humectetur, & animantia magis invalescant. Breviter: Nullum est Meteoron, quod non suam manifestam seu occultam utilitatem habeat. Hinc & Ptolemaeus inquit: Traiectiones & crinitae secundas partes habent in iudiciis, tamen ab hominibus non imprudentibus etiam earum significationes aliquo modo animadverti possunt. Deinde Theologica finalis causa est consideranda. Saepe enim Meteora praenunciant aliquid mortalibus, aut praemonent de secuturis malis, aut terrae nocent, vel utilitatem adferunt. Hinc experientia testatur, quod signa sint impressiones, & recte quidem a Philosophis annumerentur signis, sicut antea monuimus." Garcaeus, *Meteorologia*, 11v.

85. "Hinc experientia testatur, quod signa sint impressiones, & recte quidem a Philosophis annumerentur signis, sicut antea monuimus." Garcaeus, *Meteorologia*, 11v. Garcaeus's list is found at 476r–82v.

86. Meurer, *Meteorologia*, B8v. An earlier edition was published in 1592.

87. Meurer, *Meteorologia*, 269–84.

88. Gerhard, *Divine Aphroismes*, 66–80. For an account of Gerhard's views on providence, see Preus, *Post-Reformation Lutheranism*, 1:107–43.

89. Sennert, *Epitome naturalis scientiae*, 314, 320, 355. See Bodin, *Universae naturae theatrum*, 209–11; Pliny, *Historia naturalis*, 2.104.

90. Liebler, *Epitome philosophiae naturalis*, 186–222.

91. Field, "Lutheran Astrologer," 263.

92. Boner, "Origins of Comets," 100–101.

93. Translation from Field, "Lutheran Astrologer," 263.

94. Methuen, *Kepler's Tübingen*, 208; Fabbri, *Cosmologia*, 114–33. For the view that Kepler thought "teleologically," see Haase, "Kepler's Harmonies," 528–29. For the view that he emphasized efficient causes, see Kozhamthadam, *Discovery of Kepler's Laws*, 65–68.

95. Maclean, "Heterodoxy in Natural Philosophy," 7–16.

96. "Hic dicit Seneca unum bonum verbum. dicit enim, illum esse securiorem qui scit nullibi se esse securum; Ideo Philosophi sunt securissimi, quoniam ipsi sciunt haec omnia ex ordinatione naturae evenire, ideo non mirantur de illis effectibus, ut facit ignobilis vulgus; quoniam horum effectuum causas cognoscunt; et ita esse ordinatum optime a natura, Ideo sciunt se positam et ordinationem Dei castro, regno, civitate, aut in divitiis, voluptatibus, aut propriis consiliis non posse evadere; Ideo spernunt omnia temporalia, et etiam mortem, diliguntque divina et contemplantur Deum optimum maximum, quomodo hos effectus tam sapienter disposuit, et quomodo fecit: Ideo contemplantur causas rerum naturalium, et quomodo Deus illae optime et ordinatissime disposuerit." Pomponazzi, *In libros meteororum*, 80v. My translation.

CHAPTER 3: THE FERRARESE EARTHQUAKES AND THE EMPLOYMENT OF LEARNED METEOROLOGY

1. Pastor, *History of the Popes*, 18:269–70.

2. Bonfil, *Jewish Life*, 61–63.

3. Guidoboni, "Riti di calamità," 118. Guidoboni's excellent archival research forms a foundation for my investigations into the Ferrarese earthquakes.

prodigiosum habent in movendo aquas. Et cum ibi inceperit quod est transierit in signum Piscis, quod est triplicitatis aqueae et maximam habet virtutem in illa triplicitate plus quam Cancer vel Scorpius, et quod fuerit permutatio triplicitatis in illa coniunctione, tunc [enim] oportuit, ut triplicitas, quae regnabat ante hoc in mundo, esset aerea, et aer umore et convertibilitate iuvat ad multitudinem aquarum." Albertus Magnus, *De causis proprietatum elementorum*, 78.

45. For a Latin version of Avicenna's discussion of universal floods, see Alonso Alonso, "Homenaje a Avicena," 291–319.

46. Russiliano, *Apologeticus adversus cucullatos*, 152–201.

47. Nifo, *Philosopho*, 10r.

48. "I gran diluvii, per l'ordinario non possono essere universali, ma in alcuna parte solamente della terra, cosi ancho i terremoti." Maggio, *Del terremoto*, 37v. My translation.

49. "& fu similmente miracolo quel terremoto universale, che mosse tutto il mondo, nella morte del Redentore dell'humano genere GIESU CHRISTO." Maggio, *Del terremoto*, 39r–39v; emphasis in the original. My translation.

50. "Ut nihil frequentius apud literatos, imperitosque homines fuerit, quam de hac re interrogatio, atque disputatio." Galesi, *De terraemotu*, 2. My translation.

51. Galesi, *De terraemotu*, 121–23.

52. For Romei's other dialogues where he reflects courtly life and conversation, see Gundersheimer, "Trickery, Gender, and Power," 121–41.

53. Romei, *Dialogo*, 2.

54. "Ond'io, il quale nelle cose della Philosophia ho sempre piu creduto al senso, & alla ragione, ch'è qual si voglia authore, ne mai mi ho fatto conscienza di lasciare nelle mie opinioni, & Aristotile, & ogn'altro, la dove non l'authorità di questo, e di quello, ch'è solamente argomento probabile, non certo." Zuccolo, *Del terremoto*, 2–3. My translation.

55. Zuccolo, Del terremoto, 51–52.

56. Zuccolo, *Del terremoto*, 17–18; 51–52. The use of siege mines as an analogy for the causation of earthquakes did not originate with Zuccolo. An earlier example is found in Biringuccio, *Pirotechnia*, 157v.

57. "Pirro Ligorio," 109–14.

58. "È nel vero, quel che si sente hora, tutto contrario a quanto ne scrive il principe de' philosophi et di quanto osservato hanno i suoi peripatetici, che vogliono dare regula alla natura et terminato tempo, né più né meno; ciò che loro dicono è contra a quelli che si trovano nell'historie." Ligorio, *Libro de diversi terremoti*, 19. My translation.

59. "Leggesi dunque che i terremoti, in ogni tempo, in ogni età che siano stati huomini spirituali, sempre l'hanno giudicati et atribuiti alli secreti et alli penetrali delle cose mirabili d'Iddio." Ligorio, *Libro de diversi terremoti*, 2. My translation.

60. "I quali accidenti nel vero non quadrano con quello che dicono i peripatetici et l'astrologi, come cose che non appartengono a loro, ma più tosto sono cose dei segreti del mirabile Iddio et degne di recorrere più alla theologia et al senso delli divini propheti." Ligorio, *Libro de diversi terremoti*, 21. My translation. For Aristotle's argument that the existence of the universal conception of the gods is a starting point for investigating the divine, see *De caelo*, 1.3.270b5–9.

61. "Havemo preso a scrivere brevemente gli horribili segni che Iddio mand per chiamare i miseri mortali a sé, come per un bon mezzo della celeste medicina . . . Et questi vengono e sono

oltre modo terribili annuntii de li meritevoli castighi d'alchunin che sono tanto temerarii che, annegati nel pelago delle ragioni et ostinate volglie che, quasi per sintomati nel corpor terreo, negano la Providentia d'Iddio gabbati da Aristotele, da Galeno metafrasta, da Averroi et da Alexandro Aphrodiseo et dagli altri peripatetici, et tutti gli suoi accidenti del terremoto et suoi accesssi riducono alli difetti et protenti et mostri de la Natura Generante." Ligorio, *Libro de diversi terremoti*, 5. "Perciò che essi colla loro philosophia et colle loro stelle discordano da ogni buono et securo senso, per che sono alcuni che insensatamente prendono l'alta potentia di Colui donde procede il Sommo Bene." Ligorio, *Libro de diversi terremoti*, 21. My translations.

62. "Noi per la certezza, che Christiana mente habbiamo della Providentia Divina, sviandoci da i Peripatetici, affermiamo ogni cosa esser fatta dal Consiglio Divino." Sardi, *Discorsi*, 176. My translation.

63. "La causa del Terremoto variamente esplicata da i Philosophi Naturali, fu creduta incertissima, inconoscibile, & piu per coniettura, che con verità potersene ragionare." Sardi, *Discorsi*, 170. My translation.

64. Sardi, *Discorsi*, 176. Sardi most likely was following Aulus Gellius's account of Roman views of earthquakes and the gods. See *Attic Nights*, 2.28.

65. "Per la qualità della terra limosa, & tenace alcuni estimarono non essere naturalmente sottoposta a caso tanto grave la Città, & il territorio Nostro, imaginandosi che sia palustre. Io veramente replicando Il terremoto ogni afflittione, & ogni bene essere causato da Dio, che il tutto regge, & dispone, & muove, & ferma le cause naturali; non resterò di dire il territorio Ferrarese, per sua proprieta non essere palustre, poi che ne i tempi antichi fu habitato da diversi Popoli: ma dopò per varii accidenti in varii tempi fu ridotto a palude: intorno a che non mi distenderò, per che lo dimostrai sufficientemente in particolare Trattato del Sito antico del territorio, & della Città di Ferrara. Si puo ben dire, che esso territorio hebbe alzamento dal limo del Pò: & che per questa causa nelle sue parti superiori non habbia caverne grandi: onde ne i tempi andati i terremoti quasi sempre vi sono stati deboli. Ma le parti inferiori mostrano havere caverne grandi, facendo robusto, & continuante terremoto nella profondità della terra. Dissi quasi sempre i terremoti stativi deboli, perche nel MCCLXXXV a XIII di Dicembre vi fu assai robusto." Sardi, *Discorsi*, 188–89. My translation.

66. Sardi, *Discorsi*, 205–7.

67. "Ben certi che la benignità Divina castiga, & non estirpa i filgiuoli suoti." Sardi, *Discorsi*, 207. My translation.

68. Bayle, *Pensées diverses*, 7–11, 87–116.

CHAPTER 4: THE CHYMISTRY OF WEATHER

1. E.g., Frytsche, *Meteorum*, 25v.

2. Wendelen, *De caussis naturalibus pluviae*. The parenthetical citations in the next paragraph also refer to Wendelen, *De caussis naturalibus pluviae*.

3. Debus, *Chemical Philosophy*, 1:161–64.

4. On the importance of material causation in meteorology see chapter 2.

5. Aristotle, *Meteorology*, 2.9.370a26–33.

6. Eichholtz, "Formation of Metals," 141–46.

7. The authorship of the *Problemata* is still contested. If Aristotle did not write this work, it is likely that a close follower did. For the difficulties in assessing the authenticity of the *Problemata*, see Aristotle, *Problemata*, 1:xi–xviii.

8. Long, *Hellenistic Philosophy*, 157–58.

9. Freudenthal, "Problem of Cohesion," 107–16.

10. For a history of *ceraunia* in the early modern period, see Goodrum, "Questioning Thunderstones," 482–508.

11. "Saepe tamen etiam lapides fiunt ex igne cum extinguitur, quia saepe contingit corpora terrea & lapidea cadere cum coruscationibus, quia ignis fit frigidus & siccus ex sua extinctione. Et in Persia cadunt etiam in coruscationibus corpora aerea, & similia sagittis hamatis, & non possunt liquefieri, sed per ignem evaporare in fumum cogente humiditate, donec residuum sit cinis." Newman, *Summa perfectionis*, 636.

12. Albertus Magnus, *Book of Minerals*, 3.1:156–57.

13. Albertus Magnus, *Book of Minerals*, 5.4:44.

14. "Verum est, quod communis fama est quod melior expositorum omnium super libros Metheorum est Albertus." Pomponazzi, *In libros meteororum*, 77r.

15. "Videmus enim quod ventus omnia concutit et movet corpora in superficie terrae manentia. Movet enim ignem aliquando exsufflando ipsum a combustibili et extinguendo corpora accensa; aliquando autem amplius inflammando accensa corpora et faciendo quod ignis penetrat in ipsa, ita quod interius exaestuant et aduruntur. Aliquando autem sua agitatione fit ipse incensus, sicut fit in nube fulgurante, et aliquando intrans poros terrae et confricatus incendit ignem in materia sulphuris et auripigmenti." Albertus Magnus, *Meteora*, 3.2.6:134.

16. "Eodem modo videmus quod convellit aedificia et structuras magnas et arbores et saxa ingentia." Albertus Magnus, *Meteora*, 3.2.6:134.

17. Albertus Magnus, *Meteora*, 3.3.18:168.

18. "Quod autem odorem habent sulphuris omnia fulminata et foetida sunt valde, est propter vaporem sulphureum, qui spargitur ex nube; quia aqueitas ipsius vaporis mixta est cum terrestreitate calore corrumpente et adurente et perducente ad unctuositatem sulphuream. Et talis vapor procedit de nubibus aquosis in se terrestres vapores habentibus ignitos." Albertus Magnus, *Meteora*, 3.3.19:170.

19. Park, "Natural Particulars," 347–69.

20. "Hec omnia sic probabilia sunt demonstratione logica carentia. sed experientia est omnium harum discordiarum magistra." Savonarola, *De balneis*, c5r. For the importance of experience see also fol. b4r: "Nam experientia est que de his certum reddet iudicium."

21. "Unde si verba aristotilis recte contemplabimur: multum ad caliditatem balnearum antiperistasim per aquam frigidam operari non negabimus. . . . Nam videmus cavernarum aerem tempore estatis per antiparistasim infrigidari: et tempore hiemum calefieri." Savonarola, *De balneis*, b4r.

22. "Hec autem terra sic concavitatibus plena bitumen quoddam est unctuosum et solidum et compactum: ut propter eius soliditatem et unctuositatem aqua penetrare minime possit. Unde ut actum eo in loco sic congregata: eam calefieri et bullitionem et ferventiam recipere comprehendimus. Intensiorem quam caliditatem eo in loco quam in superficie terre ad quam ad nos pervenit recipiens. Nam in transitu per porositates terre frigide et ad aerem

veniens in ferventia sua sic remittitur." Savonarola, *De balneis,* b4v. See also Vermij, "Subterranean Fire," 323–47.

23. "Sicque aqua termarum a loco per quem defluit denominationem recipiens est: ut alia dicatur sulfurea a praedominio ut aqua termarum petrioli. Alia aluminosa vel salsa: ut balnea ebani in comitatu paduano." Savonarola, *De balneis,* b4v.

24. "Ferunt et inveniri alumen humidum, et hoc est sicut bitumen untuosum quod multum est cremabile deficiens in odore aliquantulum a sulfuris." Savonarola, *De balneis,* b5v.

25. "Et similiter dicamus in vulcano qui est in terra sicilie et inter magna maria. quod ille concavitates et ille ruine in quibus sunt ignes: non sunt nisi ex sulfure in ventre illius montis. ergo conflat ventus et movetur unda in illo mari coartat unda montem et calefacit propter illud sulfur. et accenditur in eo ignis et egreditur ex eo fumus etc." Savonarola, *De balneis,* b3v.

26. Biringuccio, *Pirotechnia,* 157v–58v.

27. Zuccolo, *Del terremoto,* 17–18.

28. Agricola, *De ortu & causis subterraneorum,* 33. For Theophrastus see *Historia plantarum,* 1.7, 1.11. For the Renaissance rejection of the celestial bodies as the source of subterranean fire, see Vermij, "Subterranean Fire," 336–41.

29. Agricola, *De ortu & causis subterraneorum,* 31. Strato's view on the cause of earthquakes is found in Seneca, *Naturales quaestiones,* 6.13, 2.

30. Agricola, *De ortu & causis subterraneorum,* 33.

31. Agricola, *De ortu & causis subterraneorum,* 35. My translation.

32. Agricola, *De ortu & causis subterraneorum,* 47. My translation.

33. Agricola, *De natura fossilium,* 229. My translation. Pliny discussed these minerals in bk. 36 of his *Historia naturalis.*

34. Bacci, *De thermis,* 34–35. My translation.

35. "Naptha est vocabulum Persicum & significat fluxum bituminis valde tenacis, non valde dissimilis amurcae olei, quod commixtum cum sulphure ac incensum nulla ratione nec arte restingui potest, propter eandem quam supra dixi causam. Meminit napthae Strab. lib. 15. In Babylonia etiam bitumen multum innascitur liquidum quod naphtham vocant, cuius natura est admirabilis, igni enim admotum eum corripit, & si corpus eo illitum igni admoveris deflagrat, nec aqua ullo pacto extingui potest, sed magis ardet. . . . Initio hoc loco exponenda est in genere caussa generationis ignium in subterraneis locis, ex qua deinde caussae incendiorum particularium facile intelligi possunt. Materialis caussa est terra pinguis sulphure, nitro alumine, aut alia re cremabili cum admixto unctuoso bitumine, quod et ignem cito recipere & longo tempore alere potest." Milich, *De mundi historia,* 202v–3v. My translation.

36. Nifo, *In libros meteorologicorum,* 402.

37. Nifo, *In libros meteorologicorum,* 409–11.

38. "Porro soni varii in tonitru cum fiant, in duas causas eorum varietatem Aristoteles refert, in nubis inaequalitatem, in densitate videlicet & raritate, seu crassitie & tenuitate, & in concavitates intermedias, ubi nulla densatio." Vimercati, *In quatuor libros meteorologicorum,* 121v.

39. "Sed praeclarior est similitudo, quae ex bombardis sumi potest, quibus pila exhalationis ignitae impetu excussa, sonitus maximus efficitur. Quamvis hoc interest, quod non ita bombardae, ut nubes rumpuntur." Vimercati, *In quatuor libros meteorologicorum,* 120r. My translation.

40. "Est ea vicina mari, aquis, calidis, & lutoso sulfure abundans: montes habet a septentrione & meridie, qui ad mare usque procurrunt, ubi cavernae multae & magnae caloris vim plurimam cohibent. Fuit haec regio biennio fere magnis terraemotibus agitata: ut nulla in ea superesset domus integra, nullum aedificium, quod non certam & proximam ruinam minaretur." Porzio, *De conflagratione*, 3.

41. "Quae natura raro fiunt, non una ratione oriuntur cum eorum quaedam certis careant difinitisque causis, cuiusmodi sunt quae temere & casu fiunt: alia vero non sine certis causis, tametsi ea quoque raro eveniant. huius generis sunt, luminarium dcfectus, igneae exhalationes, terraemotus." Porzio, *De conflagratione*, 4.

42. "Fumus hic in terrae antris & cavernis genitus, aut inde totuts effluit, aut totus coercetur: vel partim quidem effluit, partim coercetur. Si totus effluit, vi sua propellit aerem & generat ventos: sin coercetur totus, vel a meatibus & rimis ac arentis terrae soliditate." Porzio, *De conflagratione*, 5.

43. "Demum exhalationes hae motu velocissimo materiam in ea regione bituminosam, atque igne perustam, in terrae cavernis clausam inflammarunt, atque eam magna vi propulsam eiecerunt." Porzio, *De conflagratione*, 6. My translation.

44. "Haesitarunt quidam, an ignis ille in bitumine praeesset, an potius exhalationum motione accensus fuerit. Responsumque a nobis est, materiam quidem illam antea arsisse: argumento, quod aquae quae iuxta eam scaturiebant, calidae erant; verum exhalationum impetu, & exeundi vim & incendium maius accepisse." Porzio, *De conflagratione*, 7.

45. "Imo totius ager Puteolanus, sufureo bitumine plenus sit, ut habeat ignis ille aptam." Porzio, *De conflagratione*, 7–8.

46. Cardano, *De subtilitate*, 241. My translation.

47. "Unde nil aliud terraemotus est, quam subterraneum tonitruum est coeleste terraemotus." Della Porta, *De aeris transmutationibus*, 240. My translation. "Nil aliud terraemotum et eius species sub terra esse dicimus quam tonitrum et eius species sub coelo." Della Porta, *De aeris transmutationibus*, 239.

48. "Subterraneus ignis in bitumen et sulphur agens aliosque inflammabiles liquores, fumidam pinguemque exhalationem eliciens, quae per subterraneas speluncas fertur et dum hac, illac conclusa agitatur ab igne accenditur, quemadmodum pyrius pulvis in aeneis tormentis accensus; aut in subterraneis cuniculis, arces, castra et urbes evertit et deturbat aut multi vapor in doliis conclusus." Della Porta, *De aeris transmutationibus*, 240. My translation.

49. Della Porta, *De aeris transmutationibus*, 11.

50. Della Porta, *De aeris transmutationibus*, 30. My translation. For his discussion of natural distillation, see *De distillationibus*, 4–5. For the use of the sun, see 29–31.

51. Della Porta, *De aeris transmutationibus*, 123. My translation.

52. Della Porta, *De aeris transmutationibus*, 131. My translation.

53. Della Porta, *De refractione optices parte*, 2.22:63–64.

54. Della Porta, *De aeris transmutationibus*, 147.

55. Martin, "Alchemy," 245–62.

56. Newman, *Atoms and Alchemy*.

57. "Exhalant nonnunquam fulmina sulphuris odorem, quia spiratio, in qua ardent, multum habet sulphureae materiae . . . odoris causa erit, quod halitus ipse e sulphurea terra evocatus nativum solum redoleat." Collegium Conimbricense, *In meteorologicos*, 20.

58. "Philosophi causam quaerunt partim in reflexione multiplicationeque calidorum spirituum inflammatilium, quales sunt halitus sulphurei, bituminosi & nitrosi simul: quos antiperistatis frigidi cogit: partim in motum, quo rarefacti inflammantur, & prorumpunt cum immenso fragore." Libavius, *Commentariorum alchymiae pars prima*, 195.

59. Newman, *Atoms and Alchemy*, 83–152.

60. For Aristotle's definition of *phlogiston*, see *Meteorology*, 4.9.387b18.

61. Sennert, *Epitome*, 303.

62. Sennert, *Epitome*, 304.

63. Sennert, *Epitome*, 314–15.

64. Sennert, *Epitome*, 317.

65. AT, 1:406–8.

66. For the meaning of "mechanical" in this context, see Gabbey, "What Was 'Mechanical'?" 11–24. For Froidmont's concerns about atomism, see Garber, "Revolution," 471–86.

67. Nouhuys, *Two-Faced Janus*, 240–50.

68. Froidmont, *Meteorologicorum libri sex*, 22, 50, 71. My translation.

69. Kepler, *Harmonices mundi*, 4.7:72, 4.7:187–91.

70. "De vapore, res certior: quia fumus vapidus ex olla scandens & operculo adhaerens, in aere etiam fervido aut tepenti, in aquam revertitur, nulla ibi manifesta causa secunda praesent, quae tam cito ex vapore valeat regenerare aquae substantiam. Idem etiam ex omnibus distillationibus & sublimationibus, Chymicorum potest ostendi, ac facillimo reditu aestivi vaporis in rorem." Froidmont, *Meteorologicorum libri sex*, 26. My translation.

71. "Tarthuris, sulphuris, & infinitorum mineralium exhalationes, veterem speciem servant." Froidmont, *Meteorologicorum libri sex*, 27. My translation.

72. Froidmont, *Meteorologicorum libri sex*, 74. My translation.

73. Froidmont, *Meteorologicorum libri sex*, 85–86.

74. Froidmont, *Meteorologicorum libri sex*, 87–88.

75. Lohr, *Latin Aristotle Commentaries*, 249–50.

76. Mastri and Belluti, *Disputationes*, 311.

77. Mastri and Belluti, *Disputationes*, 224–26, 272–81.

78. Mastri and Belluti, *Disputationes*, 309.

79. "Alia opinio satis communis inter Chymicos asserit materiam proximam metallorum esse sulphur, ut semen patris, & argentum vivum, ut menstruum matris. . . . Experientia quoque fulciuntur opiniones Chymicorum, nam ex illis materiis varia conficiunt metalla & quia in venis metallorum reperiuntur sulphur, & argentum vivum. Quare suppositis his experientiis dicendum erit, quamlibet esse sufficientem materiam: at in visceribus terrae maxime generari ex humore aqueo cum subtili terreo perfecte mixto, ut dicebat Arist. & explicavit Albertus." Mastri and Belluti, *Disputationes*, 308.

80. Mastri and Belluti, *Disputationes*, 241. They based their views partially on Carafa, *De naturali concursu*.

81. "Ex omnibus aquis sive marinis, sive fluvialibus vapor extrahitur virtute Solis, & syderum. Est autem vapor halitus quidam subtilis, rarus, humidus, & calidus, sicut videmus in olla ferventi, e qua vi caloris partes aquae subtiliores per modum fumi ascendunt; de quo dubitari solet, an specie differat ab aqua, & quamvis affirmativa pars non careat probabilitate; negativa tamen est probabilior; tum quia vapores herbarum in distillatorio vase excepti per

quae sic percipi non possunt non spectabunt ad Physicam." Cabeo, *Commentaria*, 1:9. My translation.

23. "Certe pulcra sunt de Meteoris quae scripsit, neque enim sequitur vulgus philosophorum, sed quod iuxta experientias optimum judicat." Mersenne, *Correspondance*, letter 1514, 22 September 1646, 14:473. My translation. For his favorable view of this work, see also letter 1528 to André Rivet, 11 October 1646, 14:524: "Nous avons un nouveau commentaire en 2 volumes sur les *Meteores* d'Aristote fort excellent, du Jesuiste Cabeus, imprimé à Rome."

24. "& sic fortasse forma substantialis, est essentia & ratio metaphysica apud Aristotelem. Non entitas physica." Cabeo, *Commentaria*, 4:80. My translation.

25. "Non forma, & privatio, quorum alterum nihil est, alterum quid metaphysicum." Cabeo, *Commentaria*, 1:406.

26. "Forma vero physica est ille, spiritus vapidus, & subtilis, ille enim est, qui dat rei unicuique determinatum esse. Ideo enim res est talis, quia tali spiritu animatur. Ab isto est vis activa, tanta, & talis; & sicuti diversitas harum rerum sublunarium provenit a diversis istis spiritibus, qui rebus inditi sunt; ita diversitas facultatum, proprietatum, operationum, virtutum, ab iisdem prodit. Hic vero verus actus, haec vera forma, non metaphysica, mente concepta ratio, sed physicum principium facultatum." Cabeo, *Commentaria*, 3:4. My translation.

27. Cabeo does not use the world *pneuma*, but there are similarities between his form and that of the Stoics; see Armin, *Stoicorum veterum fragmenta*, §§ 439–62. For Stoic influences in the seventeenth century, see Barker, "Stoic Contributions," 135–54.

28. Poinsot, *Cursus philosophicus thomisticus*, 129.

29. Poinsot, *Cursus philosophicus thomisticus*, 129–30.

30. Newman, *Summa pefectionis*. Cabeo cited Pseudo-Geber in *Commentaria*, 4:356.

31. "Est praeterea altera pars fixa, crassa, manens, terrea, cinericia, consistens, quae nunquam evanescit." Cabeo, *Commentaria*, 1:13.

32. "Semper remanet gravis, & cadens." Cabeo, *Commentaria*, 1:113. My translation.

33. "In universum pars spiritosa dicetur forma, fixa est vera materia, humiditas est physica, & vera unio, qua forma unitur materiae." Cabeo, *Commentaria*, 1:407. My translation.

34. "A sapientioribus pars spiritosa dicitur sulphur; fixa dicitur sal, & humiditas vocatur, Mercurius." Cabeo, *Commentaria*, 1:407. "Humiditas ergo, quam, si non haberes aliam vocem, a Chymicorum vocibus vocabis Mercurium, est vinculum illud, & ligamen, quo sulphur, hoc est spiritosa, & vivida illa pars, quae in resolutione abit in auras, & subtile effluvium, coniungitur cum sale, hoc est cum parte illa fixa, stabili, & consistente, quae semper remanet, & in perfecta resolutione nunquam resolvitur." Cabeo, *Commentaria*, 1:114.

35. "Haec sunt tria vera principia physica." Cabeo, *Commentaria*, 1:406.

36. "Physicae, & reales sunt partes." Cabeo, *Commentaria*, 1:114.

37. "Probatur hoc experientia quoties quacumque ratione destruitur quodlibet mixtum semper duas istas partes observabis, alteram quae avolat, alteram quae residet, & quamvis spiritosa illa furtim se subducat, & sensum saepe eludat." Cabeo, *Commentaria*, 1:13. My translation. "Mihi sensata experientia ostendit omnia mixta constare ex duplici Physica parte quarum una est subtilis, spiritosa, quae facile in auras abit." Cabeo, *Commentaria*, 1:13.

38. "Respondeo hoc sapientissime factum; revera enim in tota arte sublimatoria, quae unica Meteora docere, duo constat esse genera effluviorum, quae ascendunt; humida, & sicca, seu aquea, & spiritosa." Cabeo, *Commentaria*, 1:113–14. My translation.

39. For Aristotle's distinction between mixture and combination, see Joachim, "Chemical Combination," 72–86. For early modern debates over the distinction of these terms, see Lüthy, "Aristotelian Watchdog," 542–61.

40. "Haec est vera generatio physica, de quae hic Philosophi, quod nimirum partibus fixis; iterum volatiles aliae separatae adiungantur, & convenienti humore adglutinentur, & haec est vera physica mixtio, & perficitur, ut constabit ex infra dicendis, concoctione illius humidi, quo partes spiritosae, cum fixis coniunguntur, & tota perfectio." Cabeo, *Commentaria*, 4:84. My translation.

41. "Tota rerum sublunarium perfecta compositio in eo consistit, ut partes sint perfecto vinculo copulatae, & quo magis coniunctae fuerint, & minus separabiles, etiam ab efficaciori agente, diceretur certe res magis perfecta, in ratione unius, & compositi; istam autem partium compositionem, seu colligationem, dixi iam saepe fieri in humido." Cabeo, *Commentaria*, 4:98. My translation.

42. "Dico ergo, ut saepe indicatum est, & non semel etiam fusius explicatum, rem aliquam corrumpi, nihil aliud esse, quam ex attenuatione humidi, quasi ex dissolutione vinculi separari spiritus, & partes subtiliores, a corporalibus: & crassas, & consistentes concidere, subtiles in auras abire." Cabeo, *Commentaria*, 4:80. My translation.

43. "Dum resolvuntur corpora, effluvia diversa ex singuli educi; & ita etiam divulsa diversam inter se retinere naturam." Cabeo, *Commentaria*, 1:114.

44. "Ut in corruptione nihil deperditur, sed quae erant unita dividuntur; in generatione nihil producitur, sed quae erant divisa uniuntur." Cabeo, *Commentaria*, 4:80.

45. Newman, "Experimental Corpuscular Theory," 294.

46. "Dum ergo partes spiritosae, ut ad rem revertar, alicuius compositi sunt commixtae cum fixo, sed humidum remanet adhuc aqueum, facile calore, avolat aqua, & ita compositum dissolvitur: unde si mixtio dicitur perfecta, & mixtum perfectum, quando non qualibet vi potest dissolvi, debet eo res redigi, ut non ita facile humidum possit avolare: debet ergo illa humiditas, quae dicebatur humiditas quanta, & aquea, quia non erat incorporata partibus componentibus, sed solum quasi localiter, & per minima permixta, bene incorporari, & uniri, interna penetratione, & identificatione, ut iam fiat humidum quale, & oleaginosum, & hic transitus fit per calorem semper, nec alio fere modo natura hoc assequitur, qui calor in principio debet esse temperatus, & moderatus; dum enim illo moderato calore attenuatur humiditas, ex una parte non avolat, quia non redditur nimis tenuis; ex alia redditur magis apta, ut spiritus illam pervadere possint, tota sua substantia: & ita illa humiditas redditur magis spiritualis, & redditur ipsa magis apta, ut intime pervadet ipsum corpus, & acquirit ingressum, dum redditur magis penetrativa, & subtilis, & ita illi intime partes fixae copulantur, & dum hoc perpetua & continuata actione fit, hoc modo corpus fit spirituale, & spiritus corpeus, & dum sic humiditas perficitur, & intimius unitur cum partibus, dicitur fieri concoctionem, proprie & vere." Cabeo, *Commentaria*, 4:99.

47. "Ista est vera doctrina peripatetica, hic expresse a Aristotele tradita; ut videas me novam non procudere [*sic*] philosophiam: sed eandem, quam Aristoteles desumpsit ab antiquis ab Aristotele repetere." Cabeo, *Commentaria*, 4:342. My translation.

48. Aristotle, *Meteorology*, 4.8.384b30–34.

49. "Puto ego Aristotelem, quicquid dicant alii Interpraetes, loqui universaliter de omnibus, non de aliquibus tantum, & concludit, quod omnia corpora similaria constant ex

quatuor corporibus, ex terra, & aqua, & ex duplici exhalatione." Cabeo, *Commentaria*, 4:342. My translation.

50. "Nam diversitas specifica corporum pendet ex diversitate harum partium, & praesertim, ex diversitate illorum halituum, quos ponit Aristoteles quamvis enim uno nomine communi vocentur halitus, haec tamen est ratio generica, & in singulis rebus diversae speciei sunt, isti halitus: & licet etiam pars fixa fortasse sit diversa; tamen istae halitus ccrte est diversus, & ex istis vere constituuntur res eiusdem vel diversae speciei. Non a rationibus abstractis per intellectum metaphysice." Cabeo, *Commentaria*, 4:346.

51. "Ex animali expirant spiritus animales, & sensitivi, per contiunuam transpirationem corporum, ut constat experientia." Cabeo, *Commentaria*, 4:81. My translation.

52. "Cum enim dissolvitur aliquod vivens, & aliquod animal, id contingit, ut dictum est, quia illi spiritus, qui coniuncti cum illo fixo, formabant illud vivens, attenuato humido separantur, & avolant; non avolant autem coniuncti, & ita non potest dici avolare animam equi, aut canis sed avolant divulsi, & distantur, ac disperduntur." Cabeo, *Commentaria*, 4:81. My translation.

53. "Hinc vides cur facilius generentur vermes ex carne putrescente aut ex simili materia; quia, scilicet, illa materia est magis referta spiritibus animalibus." Cabeo, *Commentaria*, 4:82. My translation.

54. "Nec Philosophus erit unquam physicus scientificus, solum legendo libros philosophorum, nisi ipsam naturam consideret, & experimenta sumat." Cabeo, *Commentaria*, 4:353. My translation.

55. On the division between metaphysical and empirical versions of Aristotle during the Renaissance, see Kessler, "Metaphysics or Empirical Science?" 79–101. See also Poppi, *Introduzione*, 13–44, for a characterization of Paduan Aristotelianism as privileging experience at the expense of metaphysics.

56. Dear, *Discipline and Experience*, 42–46.

57. Newman, "Art, Nature, and Experiment," 305–17; Newman, "Experimental Corpuscular Theory," 291–329.

58. "Hoc non tam ex libris, aut dictis sapientum, quam ex oculari Philosophia dicere possumus, dum enim quaero ex qua re aliquid constet, & ex quo componatur tutius fortasse erit rem intueri, quam aliquem interrogare, in hoc enim non quaerimus placita hominum, sed naturae opus. Verum quidem est, si non valemus per nos ipso hoc investigare, vel de nostris experimentis diffidimus aliorum observationibus insistendum, & experientias observandas; caeterum sensu, quo manu ducent ad naturae penetralia pervenire tutissimum est." Cabeo, *Commentaria*, 1:12. My translation.

59. Lloyd, "Experiment," 50–72. For discussions of meteorology see especially Lloyd, *Method and Problems*, 77–79, 88–91. For general observations about Aristotle on experience, see Bourgey, *Observation et experience*.

60. Newman, *Promethean Ambitions*, 242–50.

61. "Nam etiam ferrum est pressibile, & signabile, quid, quid dicat Aristoteles, ut constat experientia quotidiana, quam non debemus deferere propter dictum alicuius hominis." Cabeo, *Commentaria*, 4:393. My translation.

62. "Probavi hoc experientia in plumbo, & in aliis metallis, explorata enim per aquam, ex alibi dictis, globi plumbei magnitudine. si contundatur in laminam, & iterum exporetur eius magnitudo, constabit nihil penitus magnitudinis esse deperditum, ex mutatione figurae, per

contusionem facta; sed eadem magnitudinem conservare in lamina, quam possidebat in globo. & hoc est verum, & constat experientia, nec de hac re consulendus est Aristoteles, aut alius philosophus, sed experientia ipsa." Cabeo, *Commentaria*, 4:393. My translation.

63. "Ex quo constat ipsum mensurasse singulorum planetarum distantiam a terra, non semel, aut bis, sed frequentissime quantum requirebatur, ut posset pronuntiare, quando nam sint in maxima elongatione a terra, quando in minima, & quanta sit haec distantia." Cabeo, *Commentaria*, 1:172. My translation. See also Dear, *Discipline and Experience*, 127–29, for Cabeo's privileging of the replicability of experiences.

64. "Sed si quis prius meliora, & magis experimentis congruentia attulerit; etiam ego libenter sententiam mutabo; nec enim ista, ut mathematice demonstrata profero." Cabeo, *Philosophia magnetica*, 195. Cf. Dear, "Jesuit Mathematical Science," 172.

65. "Puto illos [Chimicos] esse veros Philosophos physicos, qui ex propriis principiis rerum naturas venantur. nec potest quis melius scire, ex quibus nam res constent, tamquam ex elementis, quam si res ipsas dissolvat, & partes componentes disiungat." Cabeo, *Commentaria*, 4:244. My translation.

66. "Hac occasione do tibi lector pulcherrimum experimentum, & eius subtexo rationem, ut videas spiritum tenuem, si inflammetur, non habere sensibilem impetum, nisi dum coniungitur cum materia aliquanto crassiore. Summe salnitrum repurgatum, & in crucibulo igne fusionis accende." Cabeo, *Commentaria*, 3:26.

67. "& hoc exacta experientia observatum est, & notavit etiam Milius, in sua Basilica chymica." Cabeo, *Commentaria*, 4:315. My translation.

68. "In homine vero, non solum sunt istae substantiae spiritosae, sed est praeterea alia substantia spiritualis, quae non est spiritosa, & vapida, sed est vere in suo esse substantiali aliquid omnino diversum a corpore, & substantia spiritualis, quae, quia unitur corpori ad complendum principium unius operationis, ad quam operationem necessario praerequiritur corpus, & quia ex sua natura est talis, & facta est ut uniatur corpori, ad perficiendam istam operationem; ideo dicitur forma informans & vere est informans, & ens incompletum, in ratione illius principii, sed haec pluribus ad libros de anima." Cabeo, *Commentaria*, 4:82. My translation.

69. Baroncini, "L'insegnamento," 182–215. For the Jesuit censuring of atomism in 1632, see Redondi, *Galileo eretico*, 304. For the continued concern about atomism later in the seventeenth century see Matton, "Note sur quelques critiques," 287–94.

70. For a transcription of the notes of some of Cabeo's censors, see Baldini, *Legem impone subactis*, 95–98. For another Jesuit of this time who espoused a similar corpuscular natural philosophy, see Roux, "Honoré Fabri," 75–94.

71. "Et tamen sunt homines, qui dictis istis acquiescant; & putent Aristotelem omnia demonstrare, & isti sunt hostes fidei nostrae. utinam enim non essent etiam nunc, qui Aristotelem magis, quam Moysem sequuntur." Cabeo, *Commentaria*, 1:397.

72. "Libenter viderem quid apponant Atheistae peripatetici, quod sit physicum, & ex sensatis deductum, & fundatum in solo lumine naturae." Cabeo, *Commentaria*, 1:419. On the meaning of the word *atheist* in early modern Europe, see Kristeller, "Myth of Renaissance Atheism," 233–43.

73. Kahn, "La condamnation," 143–93.

74. "Ad id vero Bellarminus tametsi Pontificem pronum videret, nihilominus flexit libere in sententiam adversam, dixitque periculum certius a Platone, quam ab Aristotele

NOTES TO PAGES 119-122

emanare in ingenia posse, si inter Christianos ille cathedram habeat; non quod singula eius Auctoris infecta erroribus putaret; sed quia doctrinae Catholicae magis affinis Plato, quam Aristoteles est." Fuligatti, *Vita Roberti Bellarmini*, 2.11, fol. a2r.

75. Gilson, *Le Thomisme*, 205-7.

76. Russiliano, *Apologeticus adversus cucullatos*, 174.

77. For Cremonini's contention that all he did was interpret Aristotle, see Berti, "Di Cesare Cremonini," 290: "Si diligenter observentur quae dixi videbit observator omnium esse ex Aristotelis explicatione. Quare dum Aristotelis Interpretum ago, licet mihi omnes, et quoscumque qui secus interpretantur reprobare." For Cremonini's declaration to the inquisitors that he was obliged by statute to interpret Aristotle, see Renan, *Averroès et l'averroïsme*, 479: "Non posso ne voglio retrattare le espositioni d'Aristotile, poichè l'intendo così, son pagato per dichiarlo quanto l'intendo, e nol facendo, sarei obligato alla restitutione della mercede." For the statute demanding the explication of Aristotle's text, see *Statuta almae universitatis*, 161: "Quod Doctores omnes sub poena privationis lecturae, quam legunt, teneantur, ac debeant legere, et clare exponere, ac declarare tex. Authorum, quos legere tenentur de verbo ad verbum, neque amplius aliquem expositorem." For the political and religious issues, see Piaia, "Aristotelismo," 125-45. For documents surrounding these trials, see Poppi, *Cremonini e Galilei*.

78. For the view that the human intellect is immortal, see Cremonini, *Apologia dictorum Aristotelis*, 5, 79. See Cremonini, *Expositio*, 258, for the view that God created the world. For a defense of Cremonini, see Kennedy, "Cesare Cremonini," 143-58. On seventeenth-century appropriations of Cremonini, see Bosco, "Cremonini," 249-89; Gregory, "Aristotelismo e libertinismo," 1:279-96. For an imaginative account of the influence of Cremonini's alleged libertinism, see Muir, "Why Venice?" 331-53. For a skeptical take on Cremonini's role in the history of libertinism and Averroism, see Martin, "Rethinking Renaissance Averroism," 3-19.

79. "Non ergo naturaliter, seu vi causarum naturalium, in nubibus generatur: fuit ergo ille, humor aliquis rubeus, & purpureus, quem humorem, ex similitudine vulgus hominum sanguinem appellavit." Cabeo, *Commentaria*, 1:282. My translation.

80. Collegium Conimbricense, *In meteorologicos*, 62.

81. Cabeo, *Commentaria*, 1:283.

82. Cabeo, *Commentaria*, 1:420.

83. Aristotle, *Meteorology*, 1.14.351a19-353a28.

84. The idea that Aristotle had access to the Old Testament was held by some of Cabeo's contemporaries. Some believed that his access meant his beliefs about God corresponded to Christian ones. See Liceti, *De pietate Aristotelis*, 88.

85. "Audivi ego aliquando, libros Moysis ad Aristotelis manus devenisse; qui Aristoteles legens initium. *In principio creavit Deus Caelum, & terram:* Ecce, ferunt, dixit: Iste auctor multa dicit, & nihil probat; at Moyses tunc historicum referebat, & historica erat narratio, in qua narratio, non requiritur probatio, sed simplex facti expositio. Inepte igitur Aristoteles ab historico exigit probationem. Philosophus, quae dicit, probare debet, unde optimo iure quis hic Aristoteli reponat: *Multa dicis, & nihil probas.*" Cabeo, *Commentaria*, 1:400. My translation.

86. "Non puto Aristotelem hoc attulisse, quasi philosophice; sed potius poetice, & ad elegantem loquendi formulam." Cabeo, *Commentaria*, 1:398. My translation.

87. "Habemus historias trium fere millium annorum, & sacras, & profanas; quae commemorant maria quaedam exigua, si cum Oceano conferantur: Euxinum, Ionium, Adriaticum, Caspium, Rubrum. ista maria tot seculorum decursu, non sunt mutata, non sunt exsiccata, non perierunt, nec incrementum sumpserunt, sed fere perseverant, ut erant tunc." Cabeo, *Commentaria*, 1:400. My translation.

88. "Sed dico, montes quotidie minuuntur. nec ulla fingi potest via, qua restaurentur: ergo terra haec non fuit ab eterno pro ut nunc est." Cabeo, *Commentaria*, 1:415. My translation.

89. "At est principium destruendi montes, & non est principium restaurandi. ergo status montium non est aeternus. haec mihi videtur physica demonstratio, non videntur demonstrationes, quae afferuntur ab Aristotele pro mundi aeternitate; quas in octavo physico. evidenter confutavi, nec rationes adductae ab Aristotele sunt ex per se notis, & sensatis; ut est per se notum, montes quotidie imminui, nec umquam restaurari. ergo haec imminutio non fuit ab aeterno, nec in aeternum durabit." Cabeo, *Commentaria*, 1:415. My translation. Philo Judaeus perhaps was the source of this argument; see Philo Judaeus, *De aeternitate mundi*, §§ 23–27. Philo was apparently responding to Theophrastus; see McDiarmid, "Theophrastus," 239–47.

CHAPTER 6: CAUSATION AND METHOD IN CARTESIAN METEOROLOGY

1. Gilson, "Météores cartésiens," 102–37.

2. On Descartes and Jesuit instruction at La Flèche, see Rodis-Lewis, "Un élève," 25–36; Giard, "Compagnie de Jésus," 199–225. For Descartes' studies after his time at La Flèche, see Rodis-Lewis, *Descartes*, 24–142.

3. AT, 3:185. For a discussion of Descartes' request for Jesuit textbooks, see Ariew, *Descartes*, 26.

4. Armogathe argues that Descartes was familiar with observations found in Froidmont's *Meteorologicorum libri sex*. See Armogathe, "Rainbow," 252.

5. "Compositio deinde illa corporum ex partibus diversarum figurarum pg. 159, quibus invicem tanquam uncinis cohaerescant, nimis crassa & mechanica videtur." AT, 1:406. My translation.

6. AT, 1:408. My translation. For Descartes' understanding and rejection of "real qualities," see Menn, "Greatest Stumbling Block," 182–207.

7. AT, 1:430. My translation. On this letter and the meaning of the word "mechanical" in Descartes and Froidmont, see Gabbey, "What Was 'Mechanical'?" 18; Roux, "Cartesian Mechanics," 32–35. The last two lines of the above translation are taken from Gabbey.

8. Clarke, *Descartes*, 148–55; Gilson, "Météores cartésiens," 102–3.

9. For example, Gaukroger, *Descartes' System*, 25–28.

10. Garber, "Descartes and Experiment," 94–104. For a discussion that binds these two interpretive traditions of Descartes' treatment of the rainbow, see Buchwald, "Descartes's Experimental Journey," 1–46.

11. Armogathe, "Rainbow," 249–57.

12. AT, 1:370.

13. Descartes believed that some of these explanations were based on "mathematical reasoning," such as his account of salt, although his account of salt does not use the same kinds of mathematical methods used in his discussion of the rainbow. See AT, 3:506.

14. Garber, "Descartes and Method," 43.

15. "In Dioptrica autem & Meteoris, particularia multa ex iis deduxi, quae declarant quo rationcinandi genere utar; ideoque, quamvis istam Philosophiam nondum totam ostendam, existimo tamne, ex iis quae iam dedi, facile posse intelligi qualis sit futura." AT, 7:602.

16. McMullin, "Conceptions of Science," 32–44.

17. See McMullin, "Explanation as Confirmation," 84–102.

18. AT, 6:233. My translation.

19. Garber argues that in 1637 Descartes thought that the corpuscular principles were certain; see Garber, "Descartes on Knowledge," 111–29.

20. AT, 6:315. My translation.

21. AT, 6:324.

22. AT, 6:239. My translation.

23. AT, 3:491–92. For Descartes' and Regius's diputes with Voetius, see Verbeek, *Descartes and the Dutch*, 1–33; Ruler, *Crisis of Causality*.

24. E.g., Gilson, "La critique cartésienne," 143–68; Grene, *Descartes among the Scholastics*; Garber, *Descartes' Metaphysical Physics*, 104–16.

25. AT, 6:42–43. My translation.

26. Scaliger, *Exotericarum exercitationum*, exercitatio 307, par. 29, p. 416. My translation.

27. Sennert, *Hypomnemata physica*, 46. My translation. See Newman, *Atoms and Alchemy*, 137.

28. AT, 6:43. My translation.

29. See chapter 2 for Lutherans' views on the causes of prodigious meteorological phenomena.

30. See chapter 2, where Pomponazzi is quoted. Philosophers "know that everything happens from the order of nature, and therefore they do not marvel at these effects as the unworthy vulgar do, since they recognize the causes of its effects and that it is orderly and best according to nature. Therefore they know the positioning and *ordinatio* of God." Pomponazzi, *In libros Meteororum*, 80v. My translation.

31. Daston and Park, *Wonders*, 316–17.

32. "In de philosophie moetmen altyt procederen van wonder tot gheen wonder." Beeckman, *Journal*, 2:375.

33. AT, 6:231. My translation. Cf. Werrett, "Wonders Never Cease," 129–47.

34. AT, 6:366. My translation.

35. AT, 6:232. My translation.

36. AT, 6:325. My translation.

37. AT, 6:313. My translation.

38. AT, 6:240.

39. AT, 6:321.

40. AT, 6:240–41.

41. AT, 6:239–41.

42. For the similarity of theories of matter found in Descartes and *De generatione et corruptione*, see Bohatec, *Die cartesianische Scholastik*, 138.

43. See chapter 4 for Sennert's and Froidmont's understandings of the chymical nature of the exhalations.

44. Similar explanations about the production of thunder are found in Lucretius, *De rerum natura*, 6.96–98, and Seneca, *Naturales quaestiones*, 2.17. See Gilson, "Météores cartésiens," 120.

45. AT, 6:316.

46. AT, 6:318–19.

47. AT, 6:319–20. My translation.

48. Goodrum, "Questioning Thunderstones," 482–508.

49. In 1631 he published a sustained attack on atomism; see Froidmont, *Labyrinthus*.

50. AT, 1:408.

51. AT, 1:402.

52. See chapter 4 for Froidmont's position on the persistence of the substantial forms of the elements in the exhalations.

53. AT, 1:403. My translation.

54. AT, 1: 404–5. My translation.

55. AT, 1:406. My translation.

56. Aristotle, *De generatione et corruptione*, 1.10.328b22–25.

57. AT, 6:236–37.

58. AT, 1:408. My translation.

59. AT, 1:407.

60. AT, 1:407–8.

61. "Unde halitus utrique communis nil est aliud, quam substantia levis quae e corporibus gravibus virtute caloris rarefacientis resolvitur." Froidmont, *Meteorologicorum libri sex*, 22.

62. "Terram imbre madefactam fumare, agros etiam sitientes & hiulcos, sole ictos, tenuem quemdam & velut igneum halitum expirare, oculis cernimus." Froidmont, *Meteorologicorum libri sex*, 22.

63. Froidmont, *Meteorologicorum libri sex*, 32.

64. AT, 1:417. My translation.

65. AT, 1:420–21. My translation.

66. AT, 1:421.

67. AT, 1:430.

68. Garber, "Revolution," 471–86.

69. AT, 1:455.

70. Bos, "Mercurius Cosmopolita," 4.

71. Bos, "Mercurius Cosmopolita," 10–11.

72. Verbeek, *Descartes and the Dutch*, 93.

73. "Bona dies Domini bona dies; O ho, video vos meas vertere plumas, quid vobis videtur de hoc libro, invenistisne librum ad palatum, qui rerum vobis declaret principia, causas enumeriet; Hujus ego auctor sum: nae ego sum homo doctus? quid sentitis de me? Ego complurimas frequentavi scholas, at nil nisi errores in ibi combeberam; Scholae etenim errorum sunt. At proprium tandem, genium secutus viam veritatis adtigi: Contemnsi libros quibus omnis generis scientiae exprimuntur in universum, utpote ab veritate remotos. Accinxime, ingeniumque meum eo disponi, ut in magno isto Mundi libro studerem; neglectis aliorum opinionibus, proprio iudicio adquiescere constitui." Cosmopolita, *Pentalogos*, 5–6. My translation. Compare with AT, 6:9: "Ie quittay entierement l'estude des Lettres. Et me resolvant de

ne chercher plus d'autre science que celle qui se pourroit trouver en moy-mesme, ou bien dans le grand livre du monde."

74. "Ea omnia in forma duplici spirituali & corporea, sive materiali & formali, ut sunt Elementa materialia, Elementa, formalia. Principia materialia. Principia formalia, semina matelia [*sic*], semina formalia ubi certa praedestinationum tempora exspectant." Cosmopolita, *Pentalogos*, 14–15. My translation.

75. Cosmopolita, *Pentalogos*, 35–36. My translation.

76. "Si omnes *Scholae* apud ipsum errantes sunt, cur demum, qui alias omnem devoravit sapientam, *Lugduni Batavorum*, operationibus incumbit Chymicis? ibi demum erroribus fatigabitur certus sit: non ista Chymica sunt, sed ciniflonum, Pseudochymicorum, imposturae & delirementa: Veros Chymicos *Hermeticae Medicinae* alumnos nondum adiit; Chymia enim ministra est Hermeticorum." Cosmopolita, *Pentalogos*, 53. My translation. For instruction in chymistry at Leiden during this period, see Spronsen, "Beginning of Chemistry," 329–43.

77. Cosmopolita, *Pentalogos*, 54.

78. "Cur autem figura Arcus (qui vulgo Iris dicitur) tales colores Nubi imprimantur, sola ordinatione Divina id fieri, qui foederis sui cum mortalitate notam tali Charactere exprimere voluit, quod nobis non rimari, sed admirari potius decet." Cosmopolita, *Pentalogos*, 40.

79. "Infirmitati hominum adscribendum id est, qui modum in rebus nesciunt; cum primum enim in his vel aliis quidpiam olfecerunt scientiis, persuasum sibi gerunt, digito se tangere coelos; mox chartis ignorantiam suam dispalescere finunt, quas superbis insigniunt titulis; imponunt speciosa ista facie vulgo admiranturque ab indoctis." Cosmopolita, *Pentalogos*, 5.

80. Verbeek, *Descartes and the Dutch*, 7.

81. AT, 3:460. My translation.

82. AT, 3:514. My translation.

83. AT, 3:504. My translation.

84. AT, 1:563. Gilson, *Etudes*, 183. See chaper 1 for the application of the *regressus* to meteorology.

85. AT, 8A:211; AT, 8A:253.

86. AT, 8A:203. My translation.

87. AT, 8A:328. See also Garber, "Descartes on Knowledge," 128–29.

88. See chapter 1 for Nifo's belief in the conjectural nature of the field of meteorology.

89. AT, 8A:327. Translation from Descartes, *Philosophical Writings*, 1:289.

90. Aristotle, *Meteorology*, 1.7.344a5–8. My translation.

91. AT, 8A:327. My translation.

92. AT, 5:550. My translation.

93. Verbeek, *Descartes and the Dutch*, 97.

94. Gaukroger, *Emergence*, 289–346.

95. Shapin, *Scientific Revolution*, 1–8.

EPILOGUE

1. Du Hamel, *De meteoris*, 3. My translation.

2. Du Hamel, *De meteoris*, 2–3. My translation.

3. Rohault, *Physique*, 1:126.

4. Rohault, *Physique*, 3:265–311; 3:285. My translation.

5. Rohault, *Physique*, 3:311. My translation.

6. Boyle, *Final Causes*, 87.

7. Boyle, *Final Causes*, 123.

8. Glanvill, *Plus ultra*, 60.

9. Glanvill, *Plus ultra*, 48.

10. Aristotle, *Meteorology*, 2.5.363a15.

11. Glanvill, *Plus ultra*, 48.

12. "Est ergo opinio Aristotelis terrae habitabilis partes, quae bene, & ordinatae habitantur, esse duas: nostram, quae est in cona cancri, & illam, quam antipodes habitant, quae est sub cona capricorni, hanc nostram habitabilem esse videmus. Sed illam Aristoteles habitatam esse, ut Alexan. inquit, Non per historias tradit, sed per coniecturas, eo quia Sol eodem modo se habet ad illam, sicut ad nostram. Et licet temporibus Aristotelis fortasse habitatio in illa per historias non erat cognita, ut Alexan. inquit. Tamen temporibus nostris per navigationes habitata reperitur. Asserunt enim invenisse gentes adeo versus illum polum habitantes, ut polus elevetur ad gra. 60. & sic quod Aristoteles coniecturis probavit, historia [non?] comprobatur." Nifo, *In libros meteorologicorum*, 329. My translation.

13. Collegium Conimbricense, *In quatuor libros de caelo*, cols. 402–6.

14. "Avic. autem quarta primi, & Arist. 8 physicae ausc. Dicit, quod quando ratio adversatur experientiae, tunc omittenda ratio, & standum experientiae." Pomponazzi, *Dubitationes*, 6v. For Pomponazzi's reference to Aristotle, see *Physics*, 8.3.253a32–b6.

15. "Ego dicerem quod hoc potest sciri magis per experientiam quam per rationem." Pomponazzi, *Dubitationes*, 41v.

16. For Cremonini's refusal to use Galileo's telescope, see Favaro, *Galileo*, 1:394. For Chiaramonti on supernovae and comets, see Chiaramonti, *Antitycho*; Chiaramonti, *Antiphilolaus*.

17. Sennert, *Epitome*, 216; Froidmont, *Meteorologicorum libri sex*, 112–16.

18. Cabeo, *Commentaria*, 1:15–22.

Bibliography

MANUSCRIPT SOURCES

Buridan, John. *Expositio libri meteororum.* MS, Vat. Lat. 2162. Biblioteca Apostolica Vaticana, Vatican City.

Cremonini, Cesare. *Expositio in quatuor libros Meteorologicorum.* MS, 120 (1626). Biblioteca Universitaria, Padua.

Pomponazzi, Pietro. *In libros Meteororum.* MS, R. 96 sup. Biblioteca Ambrosiana, Milan.

PRIMARY SOURCES

Agathias of Mirena. *Historiarum libri quinque.* Edited by Rudolfus Keydell. Berlin: De Gruyter, 1968.

Agricola, Georg. *De natura fossilium.* Basel: Froben, 1546.

———. *De ortu & causis subterraneorum.* Basel: Froben, 1558.

Agrippa, Camillo. *Dialogo sopra la generatione de venti, baleni, tuoni, fulgori, fiumi, laghi, valli, & montagne.* Rome: Bonfadino & Diani, 1584.

Albertus Magnus. *Book of Minerals.* Translated by Dorothy Wyckoff. Oxford: Clarendon Press, 1967.

———. *De causis proprietatum elementorum.* Edited by Paul Hossfeld. Vol. 5, bk. 2, of *Opera omnia,* edited by Bernhard Geyer and Wilhelm Kübel. Münster: Aschendorf, 1951.

———. *Meteora.* Edited by Paul Hossfeld. Vol. 6, bk. 1, of *Opera omnia,* edited by Bernhard Geyer and Wilhelm Kübel. Münster: Aschendorf, 1951.

Aristotle. *Complete Works of Aristotle: The Revised Oxford Translation.* Edited by Jonathan Barnes. 2 vols. Princeton, NJ: Princeton University Press, 1984.

———. *De generatione animalium.* Translated by A. Platt. Princeton, NJ: Princeton University Press, 1984.

———. *De generatione et corruptione.* Translated by H. H. Joachim. Princeton, NJ: Princeton University Press, 1984.

———. *Metaphysics.* Translated by W. D. Ross. Princeton, NJ: Princeton University Press, 1984.

———. *Meteorology.* Translated by H. D. P. Lee. Cambridge, MA: Harvard University Press, 1952.

———. *Problèmes.* Edited and translated by Pierre Louis. Paris: Les Belles Lettres, 1991.

Armin, Hans Friedrich August von, ed. *Stoicorum veterum fragmenta*. Leipzig: Teubner, 1921–1924.

Avempace. *Aristotle's Meteorology and Its Reception in the Arab World: With an Edition and Translation of Ibn Suwār's Treatise on Meteorological Phenomena and Ibn Bājja's Commentary on the Meteorology*. Edited and translated by Paul Lettinck. Leiden: Brill, 1999.

Averroes. *In librum Aristotelis de demonstratione maxima expositio*. Vol. 1, bk. 2, of *Aristotelis opera cum Averrois commentariis*. Venice: Giunta, 1572–1576. Reprint, Frankfurt: Minerva, 1962.

———. *In quatuor meteorologicorum Aristotelis libros*. Vol. 5 of *Aristotelis opera cum Averrois commentariis*. Venice: Giunta, 1572–1576. Reprint, Frankfurt: Minerva, 1962.

Bacci, Andrea. *De thermis*. Rome: Mascardi, 1622.

———. *Libri Tre, Ne' quali si tratta della natura, & bontà dell'acque, & specialmente del Tevere, & dell'Arno, & d'altri fonti, & fiumi del mondo*. Venice: Aldus, 1576.

Bacon, Francis. *De augmentis scientiis*. Vol. 1 of *Works*, edited by James Spedding, Robert L. Ellis and Douglas D. Heath. London: Longman, 1857–74.

Bayle, Pierre. *Pensées diverses, ecrites a l'occasion de la Cométe qui parut au mois de Decembre 1680*. Rotterdam: Leers, 1683.

Beeckman, Isaac. *Journal tenu par Isaac Beeckman de 1604 à 1634*. Edited by Cornelis de Waard. The Hague: Nijhoff, 1939–53.

Bérigard, Claude. *Circulus pisanus . . . veteri et peripatetica philosphia in Aristotelis libros de octo Physicorum; Quatuor de Coelo; Duos de Ortu & interitu; Quatuor de Meteoris, & tres de Anima*. Padua: Georgi, 1661.

Biringuccio, Vannoccio. *Pirotechnia: Li diece libri dell pirotechnia*. Venice: Padoano, 1550.

Boccadiferro, Lodovico. *Lectiones in secundum, ac tertium Meteororum Aristotelis libros*. Venice: Scoto, 1570.

———. *Lectiones super primum librum meteorologicorum Aristotelis*. Venice: Somasco, 1565.

Bodin, Jean. *Universae naturae theatrum*. Hanau: Wechel, 1605.

Borro, Girolamo. *Dialogo sul flusso e reflusso del mare*. Lucca: Busdraghi, 1561.

Boyle, Robert. *A Disquisition about the Final Causes of Natural Things*. Vol. 11 of *The Works of Robert Boyle*, edited by Michael Hunter and Edward B. Davis. London: Pickering & Chatto, 1999–2000.

———. *Origin of Forms and Qualities*. Vol. 5 of *The Works of Robert Boyle*, edited by Michael Hunter and Edward B. Davis. London: Pickering & Chatto, 1999–2000.

Breventano, Stefano. *Trattato del'origine delli venti, nomi et proprieta loro utile, et necessario a marinari, & ogni qualita di persone*. Venice: Camotio, 1571.

Buoni, Giacomo. *Terremoto dialogo*. Modena: Galdaldini, ca. 1571.

Cabeo, Niccolò. *Commentaria in libros Meteorologicorum*. Rome: Corbelletti, 1646.

———. *Philosophia magnetica*. Ferrara: Suzzi, 1629.

Carafa, Gregorio. *De naturali concursu causae primae cum secundis philosophicum opusculum . . . Addita epistola de nouissima Vesuuij conflagratione*. Naples: Savio, 1632.

Cardano, Girolamo. *De subtilitate*. Edited by Elio Nenci. Milan: Angeli, 2004.

Cesalpino, Andrea. *Peripateticarum quaestionum libri quinque*. Venice: Giunta, 1571.

Chiaramonti, Scipione. *Antiphilolaus in quo Philolao rediuiuo de motu terrae, & soli, ac fixarum quiete repugnatur: Rationesque eius, quas ipse pro demonstrationibus affert, fallaces detegun-*

tur; Insuper positio eadem de re Copernici confutatur, & Galilaei defensiones reijciuntur. Cesena: Neri, 1643.

———. *Antitycho in quo contra Tychonem Brahe, & nonnullos alios rationibus eorum ex opticis, & geometricis principijs solutis demonstratur cometas esse sublunares non coelestes.* Venice: Deuchino, 1621.

———. *In Aristotelem de iride, de corona, de pareliis, et virgis commentaria.* Venice: Banca, 1668.

Collegium Conimbricense. *In Meteorologicos Aristotelis Stagiritae.* Cologne: Zetzner, 1600.

———. *In quatuor libros de Caelo.* Lyon: Cardon, 1608.

Cosmopolita, Mercurius. *Pentalogos in libri cujusdam Gallico idiomate evulgati quatuor discursuum, de la méthode; dioptrique; météorique; & géométrique.* The Hague: Spryt, 1640.

Cremonini, Cesare. *Apologia dictorum Aristotelis: De quinta caeli substantia.* Venice: Meietti, 1616.

———. *Disputatio de coelo.* Venice: Baglioni, 1613.

Della Porta, Giovan Battista. *De aeris transmutationibus.* Rome: Zanetti, 1610.

———. *De aeris transmutationibus.* Edited by Alfonso Paolella. Naples: Edizioni Scientifiche Italiane, 2000.

———. *De distillationibus libri ix.* Strasbourg: Zetzner, 1609.

———. *De refractione optices parte.* Naples: Salviani, 1593.

Descartes, René. *Oeuvres.* Edited by Charles Adam and Paul Tannery. 11 vols. Paris: Vrin, 1982–91.

———. *The Philosophical Writings of Descartes.* Translated by John Cottingham. Cambridge: Cambridge University Press, 1985.

Diderot, Denis, and Jean le Rond d'Alembert. *Encyclopédie ou Dictionnaire raisonné des sciences, des arts et des métiers.* 17 vols. Paris: Briasson, 1751–65.

Du Hamel, Jean Baptiste. *De meteoris et fossilibus libri duo.* Paris: Lamy, 1660.

Finé, Oronce. *Protomathesis.* Paris: Morre & Pierre, 1532.

Froidmont, Libert. *Labyrinthus, sive de Compositione continui.* Antwerp: Plantiniana, 1631.

———. *Meteorologicorum libri sex.* London: Tyler, 1656.

Frytsche, Marcus. *Meteorum, hoc est Impressionum aerearum et mirabilium naturae operum.* Wittenberg: Cratoniana, 1598.

Fuligatti, Giacomo. *Vita Roberti Bellarmini politiani societate Iesu.* Liège: Ouwerx & Streel, 1626.

Gaetano of Thiene. *In quattuor Aristotelis metheororum libros expositio.* Padua: Maufer, 1476.

Galesi, Agostino. *De terraemotu liber.* Bologna: Benacci, 1571.

Garcaeus, Johannes. *Meteorologia.* Wittenberg: n.p., 1584.

Gerhard, Johann. *A Golden Chaine of Divine Aphroismes.* Translated by Ralph Winterton. Cambridge: Printers to the Universitie, 1632.

Glanvill, Joseph. *Plus ultra; or, the Progress and Advancement of Knowledge since the Days of Aristotle.* London: Collins, 1668.

Gorzoni, Giuseppe. *Istoria del collegio di Mantova della compagnia di Gesù.* Edited by Antonella Bilotto and Flavio Rurale. Mantua: Arcari, 1997.

Gozze, Nicolò Vito de. *Discorsi sopra le Metheore d'Aristotele, Ridotti in dialogo & divisi in quattro giornate.* Venice: Ziletti, 1584.

Kepler, Johannes. *Harmonices mundi.* Linz: Planck, 1619.

Libavius, Andreas. *Commentariorum alchymiae pars prima.* Frankfurt: Saur, 1606.

Liceti, Fortunio. *De pietate Aristotelis erga Deum & homines libri duo.* Udine: Schiratti, 1645.

Liebler, Georg. *Epitome philosophiae naturalis, ex Aristotelis summi philosophi libris ita excerpta.* Basel: Oporinus, 1561.

Ligorio, Pirro. *Libro de diversi terremoti.* Edited by Emanuela Guidoboni. Rome: De Luca, 2005.

Longiano, Sebastiano Fausto da. *Meteorologia, cioè discorso de le impressioni humide & secche.* Venice: Navo, 1542.

Luther, Martin. *D. Martin Luthers Werke kritische Gesamtausgabe.* Weimar: Böhlau, 1883.

Maggio, Lucio. *Del terremoto dialogo del Signor Lucio Maggio gentil'huomo bolognese.* Bologna: Benacci, 1571.

Malagola, Carlo, ed. *Statuti delle università e dei collegi dello studio bolognese.* Bologna: Zanichelli, 1888.

Mastri, Bartolomeo, and Bonaventura Belluti. *Disputationes in libros de celo et Metheoris.* Venice: Ginami, 1640.

Melanchthon, Phillip. *Initia doctrinae physicae.* Wittenberg: Seitz, 1553.

Mersenne, Marin. *Correspondance du P. Marin Mersenne, religieux minime.* Edited by Paul Tannery and Cornelis de Waard. 17 vols. Paris: Beauchesne, 1932–1988.

Meurer, Wolfgang. *Meteorologia.* Leipzig: Grossius, 1606.

Milich, Jakob. *Liber II: C. Plinii de mundi historia, cum commentariis.* Frankfurt: Brubach, 1543.

Nifo, Agostino. *De nostrarum calamitatum causis liber ad Oliverium Carafam.* Venice: Locatello, 1505.

———. *De verissimis temporum signis commentariolus.* Venice: Scoto, 1540.

———. *Expositio super octo Aristotelis Stagiritae libros de physico auditu.* Venice: Giunta, 1552.

———. *In libros meteorologicorum, in librum de Mistis, sive Quartum Meteororum, ab antiquis nuncupatum & ordinatum.* Venice: Scoto, 1560.

———. *Philosopho contra il falso giudicio che debba vegnir il diluvio: Per la congioncione de tutti gli pianeti in Pesci qual sera per tutto lo anno, 1524.* Venice: Da Sabio, 1521.

Olympiodorus. *In Aristotelis meteora commentaria.* Vol. 12, bk. 2, of *Commentaria in Aristotelem graeca,* edited by Wilhelm Stüve. Berlin: Reimer, 1883–1909.

Petrarca, Francesco. *Invectives.* Translated by David Marsh. Cambridge, MA: Harvard University Press, 2003.

Philo Judaeus. *De aeternitate mundi.* Edited by Franciscus Cumont. Berlin: Reimeri, 1891.

Piccolomini, Francesco. *Librorum ad scientiam de natura attinentium pars quarta: In qua Meterologica explicantur, & connexa cum eis.* Venice: Franceschini, 1596.

Poinsot, John. *Cursus philosophicus thomisticus: Tomus tertius philosophia naturalis.* Paris: Vivès, 1883.

Pomponazzi, Pietro. *De naturalium effectuum causis sive de incantationibus.* Basel: Petri, 1567.

———. *Dubitationes in quartum meteorologicorum.* Venice: Franceschini, 1563.

Pontano, Giovanni. *I poemi astrologici.* Edited by Mauro De Nichilo. Bari: Dedalo, 1975.

Porrata Spinola, Giovanni Francesco. *Discorso sopra l'origine de' fuochi gettati Dal Monte Vesevo.* Lecce: Micheli, 1632.

Porzio, Simone. *De conflagratione agri Puteolani . . . epistola.* Florence: Torrentino, 1551.

———. *De humana mente, disputatio.* Florence: Torrentino, 1551.

———. *De rerum naturalium principiis.* Naples: Cancer, 1553.

Rangoni, Tommaso. *De vera diluvii pronosticatione anni, 1524*. Rome, 1522.

Rao, Cesare. *I Meteori*. Venice: Varisco, 1582.

Rohault, Jacques. *Traité de physique*. Paris: Savreux, 1671.

Romei, Annibale. *Dialogo del Conte Annibale Romei Gentil'huomo Ferrarese*. Ferrara: Baldini, 1587.

Russiliano, Tiberio. *Apologeticus adversus cucullatos*. Edited by Luigi De Franco. Cosenza: Periferia, 1991.

Sagri, Nicolò. *Ragionamenti sopra le varietà de i flussi et riflussi del mare oceano occidentale*. Venice: Guerra, 1574.

Sardi, Alessandro. *Discorsi*. Venice: Gioliti, 1587.

Savonarola, Michele. *De balneis*. Ferrara: Belfort, 1485.

Scaliger, Julius Caesar. *Exotericarum exercitationum liber xv*. Paris: Vascosan, 1557.

Schegk, Jacob. *In reliquos naturalium Aristotelis libros commentaria*. Basel: Hevagius, 1550.

Schreiner, Johannes. *Disputata meteorologica de pluvia*. Leipzig: Kirsch, 1626.

Seneca. *Naturales quaestiones*. Translated by T. H. Corcoran. Cambridge, MA: Harvard University Press, 1971.

Sennert, Daniel. *Epitome naturalis scientiae*. Wittenberg: Helwig, 1633.

———. *Hypomnemata physica*. Frankfurt: Schleichius, 1636.

Speroni, Sperone. *Opere*. 2 vols. Venice: Occhi, 1740. Reprint, Rome: Vecchiarelli, 1989.

Spinoza, Benedictus. *The Collected Works of Spinoza*. Translated and edited by Edwin M. Curley. Princeton, NJ: Princeton University Press, 1985.

Statuta almae universitatis dd: Philosophorum, et medicorum cognomento artistarum Patavini Gymnasii. Padua: Pasquato, 1607.

Taurellus, Nicolaus. *Alpes caesae: Hoc est, Andr; Caesalpini Itali, monstrosa & superba dogmata, discussa & excussa*. Frankfurt: Palthenius, 1597.

Theophrastus. *On Weather Signs*. Translated and edited by David Sider and Carl Wolfram Brunschön. Leiden: Brill, 2007.

Thomas Aquinas. *In Aristotelis libros De caelo et mundo, de generatione et corruptione, meteorologicorum expositio*. Edited by Raimondo M. Spiazzi. Rome: Marietti, 1952.

———. *In Aristotelis libros Peri hermeneias et Posteriorum analyticorum expositio*. Edited by Raimondo M. Spiazzi. Rome: Marietti, 1964.

Titelmans, Frans. *Compendium philosophiae naturalis, seu De consideratione rerum naturalium, earumque ad suum creatorem reductione, libri xii*. Lyon: Rouille, 1558.

Trevisi, Antonio. *Sopra la inondatione del fiume*. Rome: Baldo, 1560.

Vieri, Francesco de'. *Trattato delle metheore*. Florence: Marescotti, 1573.

Vimercati, Francesco. *In quatuor libros Aristotelis Meteorologicorum commentarii*. Venice: Vascosanum, 1565.

Wendelen, Govaart. *De caussis naturalibus pluviae purpureae Bruxellensis, clarorum virorum*. London: Tyler, 1656.

Wildenberg, Hieronymus. *Totius naturalis physicae in physicam Aristotelis epitome*. Basel: Oporinus, 1548.

Zabarella, Giacomo. *De rebus naturalibus libri xxx*. Venice: Meietti, 1590.

Zuccolo, Gregorio. *Del terremoto, trattato*. Bologna: Benacci, 1571.

Zuccolo, Vitale. *Dialogo delle cose meteorologiche*. Venice: Meietti, 1590.

SECONDARY SOURCES

Alonso Alonso, Manuel. "Homenaje a Avicena en su milenario: Las traducciones de Juan González de Burgos y Salomón." *Al-Andalus* 14 (1949): 291–319.

Ariew, Roger. *Descartes and the Last Scholastics.* Ithaca, NY: Cornell University Press, 1999.

Armogathe, Jean-Robert. "The Rainbow: A Privileged Epistemological Model." In *Descartes' Natural Philosophy,* edited by Stephen Gaukroger, John Schuster, and John Sutton, 249–57. London: Routledge, 2002.

Baldini, Ugo. *Legem impone subactis: Studi su filosofia e scienza dei Gesuiti in Italia 1540–1632.* Rome: Bulzoni, 1992.

———. "The Roman Inquisition's Condemnation of Astrology: Antecedents, Reasons and Consequences." In *Church, Censorship and Culture in Early Modern Italy,* edited by Gigliola Fragnito and Adrian Belton, 79–110. Cambridge: Cambridge University Press, 2001.

———. *Saggi sulla cultura della Compagnia di Gesù (secoli XVI–XVIII).* Padua: Cleup, 2000.

Baldwin, Martha R. "Magnetism and the Anti-Copernican Polemic." *Journal for the History of Astronomy* 16 (1985): 155–74.

Bandini, Angelo Maria, ed. *Clarissimorum Italorum et Germanorum epistolae ad Petrum Victorium.* Florence: n.p., 1758–1760.

Barker, Peter. "Stoic Contributions to Early Modern Science." In *Atoms, Pneuma, Tranquillity: Epicurean and Stoic Themes in European Thought,* edited by Margaret J. Osler, 135–54. Cambridge: Cambridge University Press, 1991.

Barker, Peter, and Bernard R. Goldstein. "Realism and Instrumentalism in Sixteenth-Century Astronomy." *Perspectives on Science* 6 (1998): 232–58

Barnes, Robin Bruce. *Prophecy and Gnosis: Apocalypticism in the Wake of the Lutheran Reformation.* Stanford, CA: Stanford University Press, 1988.

Baroncini, Gabriele. "L'insegnamento della filosofia naturale nei collegi italiani dei Gesuiti (1610–1670): Un esempio di nuovo aristotelismo." In *La "Ratio Studiorum": Modelli culturali e pratiche educative dei Gesuiti in Italia tra Cinque e Seicento,* edited by Gian Paolo Brizzi, 163–215. Rome: Bulzoni, 1981.

Bausi, Francesco. "Medicina e filosofia nelle *Invective contra medicum* (Petrarca, L'Averroismo, L'eternità del mondo)." In *Petrarca e la medicina,* edited by Monica Berté, Vincenzo Fera, and Tiziana Pesenti, 19–52. Messina: Centro interdipartimentale di studi umanistici, 2006.

Belo, Catarina. *Chance and Determinism in Avicenna and Averroes.* Leiden: Brill, 2007.

Berti, Domenico. "Di Cesare Cremonini e della sua controversia con l'Inquisizione di Padova e di Roma." *Atti della Reale Accademia dei Lincei,* 3rd ser., 2 (1877/1878): 273–99.

Bohatec, Josef. *Die cartesianische Scholastik in der Philosophie und reformierten Dogmatik des 17. Jahrhunderts.* Leipzig: Deichert, 1912.

Boner, Patrick J. "Kepler on the Origins of Comets: Applying Earthly Knowledge to Celestial Events." *Nuncius* 21 (2006): 31–47.

Bonfil, Roberto. *Jewish Life in Renaissance Italy.* Translated by Anthony Oldcorn. Berkeley: University of California Press, 1994.

Bos, Erik-Jan. "Mercurius Cosmopolita: The Hermetic Response to Descartes." Paper presented at the annual meeting of the British Society for the History of Science, March 2007.

Bosco, Domenico. "Cremonini e le origini del libertinismo." In *Cesare Cremonini (1550–1631): Il suo pensiero e il suo tempo, Convegno di studi Cento, 1984,* 249–89. Cento: Centro studi Girolamo Baruffaldi, 1990.

Bourgey, Louis. *Observation et experience chez Aristote.* Paris: Vrin, 1955.

Buchwald, Jed Z. "Descartes's Experimental Journey past the Prism and through the Invisible World to the Rainbow." *Annals of Science* 65 (2008): 1–46.

Cassirer, Ernst. *Das Erkenntnisproblem in der Philosophie und Wissenschaft der neueren Zeit.* 3rd ed. Berlin: Cassirer, 1922.

Clarke, Desmond M. *Descartes: A Biography.* Cambridge: Cambridge University Press, 2006.

Clucas, Stephen. "Galileo, Bruno and the Rhetoric of Dialogue in Seventeenth-Century Natural Philosophy." *History of Science* 46 (2008): 405–29.

Cook, Harold J. "The Cutting Edge of a Revolution? Medicine and Natural History near the Shores of the North Sea." In *Renaissance and Revolution,* edited by Judith Veronica Field and Frank A. J. L. James, 45–95. Cambridge: Cambridge University Press, 1993.

———. *Matters of Exchange: Commerce, Medicine, and Science in the Dutch Golden Age.* New Haven, CT: Yale University Press, 2007.

Cox, Virginia. *The Renaissance Dialogue: Literary Dialogue in Its Social and Political Contexts, Castiglione to Galileo.* Cambridge: Cambridge University Press, 1992.

Daston, Lorraine, and Katharine Park. *Wonders and the Order of Nature, 1150–1750.* New York: Zone, 1998.

Dear, Peter. *Discipline and Experience: The Mathematical Way in the Scientific Revolution.* Chicago: Chicago University Press, 1995.

———. "Jesuit Mathematical Science and the Reconstitution of Experience in the Early Seventeenth Century." *Studies in the History and Philosophy of Science* 18 (1987): 133–175.

———. *Revolutionizing the Sciences: European Knowledge and Its Ambitions, 1500–1700.* Princeton, NJ: Princeton University Press, 2001.

Debus, Allen G. *The Chemical Philosophy: Paracelsian Science and Medicine in the Sixteenth and Seventeenth Centuries.* 2 vols. New York: Science History Publications, 1977.

Des Chene, Dennis. *Physiologia: Natural Philosophy in Late Aristotelian and Cartesian Thought.* Ithaca, NY: Cornell University Press, 1996.

Ducos, Joëlle. *La météorologie en français au Moyen Age (XIIIe–XIVe siècles).* Paris: Honoré Champion, 1998.

Duhem, Pierre. *Sozein ta phainomena, essai sur la notion de théorie physique de Platon à Galilée.* Paris: Hermann, 1908.

Eichholtz, D. E. "Aristotle's Theory of the Formation of Metals and Minerals." *Classical Quarterly* 43 (1949): 141–46.

Fabbri, Natacha. *Cosmologia e armonia in Kepler e Mersenne: Contrappunto a due voci sul tema dell'Harmonice mundi.* Florence: Olschki, 2003.

Facciolati, Jacobo. *Fasti gymnasii patavini.* Padua: Manfrè, 1757.

Favaro, Antonio. *Galileo e lo studio di Padova.* 2 vols. Florence: Le Monnier, 1883.

Field, J. V. "A Lutheran Astrologer: Johannes Kepler." *Archive for History of Exact Sciences* 31 (1984): 189–272.

Finocchiaro, Maurice. *The Galileo Affair: A Documentary History.* Berkeley: University of California Press, 1989.

Fiorentino, Francesco. *Studi e ritratti della rinascenza*. Bari: Laterza, 1911.

Franklin, James. *The Science of Conjecture*. Baltimore: Johns Hopkins University Press, 2001.

Freeland, Cynthia. "Scientific Explanation and Empirical Data in Aristotle's *Meteorology*." *Oxford Studies in Ancient Philosophy* 8 (1990): 62–107.

Freudenthal, Gad. "The Problem of Cohesion between Alchemy and Natural Philosophy: From Unctuous Moisture to Phlogiston." In *Alchemy Revisited*, edited by Z. R. W. M. von Martels, 107–16. Leiden: Brill, 1990.

Furley, David. "The Rainfall Example in *Physics* II.8." In *Cosmic Problems*, 115–20. Cambridge: Cambridge University Press, 1989.

Gabbey, Alan. "What Was 'Mechanical' about 'The Mechanical Philosophy'?" In *The Reception of the Galilean Science of Motion in Seventeenth-Century Europe*, edited by Carla Rita Palmerino and J. M. M. H. Thijssen, 11–24. Dordrecht: Kluwer, 2004.

Garber, Daniel. "Descartes and Experiment in the *Discourse* and the *Essays*." In Garber, *Descartes Embodied*, 85–110.

———. "Descartes and Method in 1637." In Garber, *Descartes Embodied*, 33–51.

———. *Descartes Embodied: Reading Cartesian Philosophy through Cartesian Science*. Cambridge: Cambridge University Press, 2001.

———. *Descartes' Metaphysical Physics*. Chicago: Chicago University Press, 1992.

———. "Descartes on Knowledge and Certainty: From the *Discours* to the *Principia*." In Garber, *Descartes Embodied*, 111–29.

———. "Descartes, the Aristotelians, and the Revolution That Did Not Happen in 1637." *Monist* 71 (1988): 471–86.

Gaukroger, Stephen. *Descartes' System of Natural Philosophy*. Cambridge: Cambridge University Press, 2002.

———. *The Emergence of a Scientific Culture*. Oxford: Oxford University Press, 2006.

Genuth, Sara Schechner. *Comets, Popular Culture, and the Birth of Modern Cosmology*. Princeton, NJ: Princeton University Press, 1997.

Giard, Luce. "Sur la compagnie de Jésus et ses collèges vers 1600." In *René Descartes (1596–1650): Célébrations nationales du quadricentenaire de sa naissance*, 199–225. La Flèche: Prytanée national militaire, 1997.

Gilson, Etienne. "La critique cartésienne des formes substantielles." In Gilson, *Etudes sur le rôle de la pensée médiévale*, 143–68.

———. *Etudes sur le rôle de la pensée médiévale dans la formation du système cartésien*. Paris: Vrin, 1951.

———. "Météores cartésiens et météores scolastiques." In Gilson, *Etudes sur le rôle de la pensée médiévale*, 102–37.

———. *Le Thomisme: Introduction à la philosophie de Saint Thomas D'Aquin*. 4th ed. Paris: Vrin, 1942.

Ginzburg, Carlo. *The Night Battles: Witchcraft and Agrarian Cults in the Sixteenth and Seventeenth Centuries*. Translated by Anne Tedeschi and John Tedeschi. Baltimore: Johns Hopkins University Press, 1992.

Girardi, Raffaele. *La società del dialogo: Retorica e ideologia nella letteratura conviviale del Cinquecento*. Bari: Adriatica, 1989.

Giustiniani, Lorenzo, ed. *I tre rarissimi opusculi di Simone Porzio di Girolamo Borgia e di Marcantonio delli Falconi.* Naples: Marotta, 1817.

Goodrum, Matthew R. "Questioning Thunderstones and Arrowheads: The Problem of Recognizing and Interpreting Stone Artifacts in the Seventeenth Century." *Early Science and Medicine* 5 (2008): 482–508.

Graiff, Franco. "I prodigi e l'astrologia nei commenti di Pietro Pomponazzi al *De caelo,* alla *Meteora* e al *De generatione.*" *Medioevo* 2 (1976): 331–61.

Gregory, Tullio. "Aristotelismo e libertinismo." In *Aristotelismo veneto e scienza moderna,* edited by Luigi Olivieri, 1:279–96. Padua: Antenore, 1983.

Grendler, Paul F. *The Universities of the Italian Renaissance.* Baltimore: Johns Hopkins University Press, 2002.

Grene, Marjorie. *Descartes among the Scholastics.* Milwaukee: Marquette University Press, 1991.

Guerlac, Henry. "The Poets' Nitre: Studies in the Chemistry of John Mayow." *Isis* 45 (1954): 243–55.

Guidoboni, Emanuela. "Riti di calamità: Terremoti a Ferrara nel 1570–1574." *Quaderni Storici,* n.s., 55 (1984): 107–35.

Gundersheimer, Werner L. "Trickery, Gender, and Power: The *Discorsi,* of Annibale Romei." In *Urban Life the Renaissance,* edited by Susan Zimmerman and Ronald F. E. Weissman, 121–41. Newark: University of Delaware Press, 1989.

Haas, Frans A. J. de. "Mixture in Philoponus. An Encounter with a Third Kind of Potentiality." In *The Commentary Tradition on Aristotle's De generatione et corruptione,* edited by J. M. M. H. Thijssen and H. A. G. Braakhuis, 21–46. Turnhout: Brepols, 1999.

Haase, Rudolf. "Kepler's Harmonies, between Pansophia and Mathesis Universalis." In *Kepler: Four Hundred Years,* edited by Arthur Beer and Peter Beer, 519–33. Oxford: Pergamon, 1975.

Hacking, Ian. *The Emergence of Probability: A Philosophical Study of Early Ideas about Probability, Induction and Statistical Inference.* Cambridge: Cambridge University Press, 1975.

Hankins, James. Introduction to *The Cambridge Companion to Renaissance Philosophy,* edited by James Hankins, 1–10. Cambridge: Cambridge University Press, 2007.

Hasse, Dag Nikolaus. "Spontaneous Generation and the Ontology of Forms in Greek, Arabic and Medieval Latin Sources." In *Classical Arabic Philosophy: Sources and Reception,* edited by Peter Adamson, 150–75. London: Warburg Institute, 2007.

Hellmann, Gustav. *Die Meteorologie in den deutschen Flugschriften und Flugblattern des XVI. Jahrhunderts: ein Beitrag zur Geschichte der Meteorologie.* Berlin: Akademie der Wissenschaften, 1921.

Hellyer, Marcus. *Catholic Physics: Jesuit Natural Philosophy in Early Modern Germany.* Notre Dame, IN: University of Notre Dame Press, 2005.

Heninger, S. K. *A Handbook of Renaissance Meteorology, with Particular Reference to Elizabethan and Jacobean Literature.* Durham, NC: Duke University Press, 1960.

Ingegno, A. "Niccolò Cabeo." In *Dizionario biografico degli italiani,* 15:686–88. Rome: Instituto della Enciclopedia italiana, 1960.

Jacob, James R. *The Scientific Revolution: Aspirations and Achievements, 1500–1700.* Atlantic Highlands, NJ: Humanities Press, 1998.

Janković, Vladimir. *Reading the Skies: A Cultural History of English Weather, 1650–1820.* Chicago: University of Chicago Press, 2000.

Jenks, Stuart. "Astrometeorology in the Middle Ages." *Isis* 74 (1983): 185–210.

Joachim, Harold H. "Aristotle's Conception of Chemical Combination." *Journal of Philology* 29 (1904): 72–86.

Johnson, Monte Ransome. *Aristotle on Teleology*. Oxford: Oxford University Press, 2005.

Kahn, Didier. "La condamnation des thèses d'Antoine de Villon et Étienne de Clave contre Aristote, Paracelse et les 'cabalistes' (1624)." *Revue d'histoire des sciences* 55 (2002): 143–93.

Kennedy, Leonard A. "Cesare Cremonini and the Immortality of the Human Soul." *Vivarium* 18 (1980): 143–58.

Kessler, Eckhard. "The Intellective Soul." In *The Cambridge History of Renaissance Philosophy*, ed. Charles B. Schmitt and Quentin Skinner, 486–534. Cambridge: Cambridge University Press, 1988.

———. "The Lefèvre Enterprise." In *Philosophy in the Sixteenth and Seventeenth Centuries: Conversations with Aristotle*, edited by Constance Blackwell and Sachiko Kusukawa, 1–22. Aldershot: Ashgate, 1999.

———. "Metaphysics or Empirical Science? The Two Faces of Aristotelian Natural Philosophy in the Sixteenth Century." In *Renaissance Readings of the Corpus Aristotelicum*, edited by Marianne Pade, 79–101. Copenhagen: Museum Tusculanum, 2001.

Koyrè, Alexandre. *Newtonian Studies*. Cambridge, MA: Harvard University Press, 1965.

Kozhamthadam, Job. *The Discovery of Kepler's Laws: The Interaction of Science, Philosophy, and Religion*. Notre Dame, IN: Notre Dame University Press, 1994.

Kristeller, Paul O. "The Myth of Renaissance Atheism and the French Tradition of Free Thought." *Journal of the History of Philosophy* 6 (1968): 233–43.

———. *La tradizione aristotelica nel Rinascimento*. Padua: Antenore, 1962.

Kusukawa, Sachiko. *The Transformation of Natural Philosophy: The Case of Philip Melanchthon*. Cambridge: Cambridge University Press, 1995.

De Leemans, Pieter, and Michèle Goyens, eds. *Aristotle's "Problemata" in Different Times and Tongues*. Louvain: Leuven University Press, 2006.

Leijenhorst, C. H. *The Mechanisation of Aristotelianism: The Late Aristotelian Setting of Thomas Hobbes' Natural Philosophy*. Leiden: Brill, 2002.

Leppin, Volker. *Antichrist und Jüngster Tag: Flugschriften: das Profil apokalyptischer Flugschriftenpublizistik im deutschen Luthertum 1548–1618*. Gütersloh: Gütersloher, 1999.

Librandi, Rita, ed. *La metaura d'Aristotile: Volgarizzamento fiorentino anonimo del XIV secolo*. Naples: Liguori, 1995.

Lines, David A. "Natural Philosophy in Renaissance Italy: The University of Bologna and the Beginnings of Specialization." *Early Science and Medicine* 6 (2001): 267–320.

———. "Teaching Physics in Louvain and Bologna." In *Scholarly Knowledge: Textbooks in Early Modern Europe*, edited by Emidio Campi et al., 183–203. Geneva: Librarie Droz, 2008.

Lloyd, G. E. R. "Experiment in Early Greek Philosophy and Medicine." *Proceedings of the Cambridge Philological Society*, n.s., 10 (1964): 50–72.

———. *Method and Problems in Greek Science: Selected Papers*, 70–99. Cambridge: Cambridge University Press, 1991.

Lohr, Charles H. "Les jésuites et l'aristotélisme du XVIe siècle." In *Les jésuites à la Renaissance: Système éducatif et production du savoir*, edited by Luce Giard, 79–91. Paris: Presses Universitaires de France, 1995.

————. *Renaissance Authors*. Vol. 2 of *Latin Aristotle Commentaries*. Florence: Olschki, 1988.

Long, Anthony. *Hellenistic Philosophy*. 2nd ed. Berkeley: University of California Press, 1986.

Lüthy, Christoph. "An Aristotelian Watchdog as Avant-Garde Physicist: Julius Caesar Scaliger." *Monist* 84 (2001): 542–61.

Maclean, Ian. "Heterodoxy in Natural Philosophy and Medicine: Pietro Pomponazzi, Guglielmo Gratarolo, Girolamo Cardano." In *Heterodoxy in Early Modern Science and Religion*, edited by John Brooke and Ian Maclean, 1–29. Oxford: Oxford University Press, 2005.

————. *Logic, Signs and Nature*. Cambridge: Cambridge University Press, 2002.

Marangon, Paolo. "Aristotelismo e cartesianesimo: Filosofia accademica e libertini." In vol. 4, bk. 2, of *Storia della cultura veneta*, edited by Girolamo Arnaldi and Manlio Pastore Stocchi, 95–114. Vicenza: N. Pozza, 1984.

Marsh, David. *The Quattrocento Dialogue: Classical Tradition and Humanist Innovation*. Cambridge, MA: Harvard University Press, 1980.

Martin, Craig. "Alchemy and the Renaissance Commentary Tradition on *Meteorologica* IV." *Ambix* 51 (2004): 245–262.

————. "Experience of the New World and Aristotelian Revisions of the Earth's Climates during the Renaissance." *History of Meteorology* 3 (2006): 1–15.

————. "Rethinking Renaissance Averroism." *Intellectual History Review* 17 (2007): 3–19.

————. "With Aristotelians like These, Who Needs Anti-Aristotelians? Chymical Corpuscular Matter Theory in Niccolò Cabeo's Meteorology." *Early Science and Medicine* 11 (2006): 135–61.

Matton, Sylvain. "Note sur quelques critiques oubliées de l'atomisme: À propos de la transsubstantiation eucharistique." *Revue d'histoire des sciences* 55 (2002): 287–94.

McDiarmid, John Brodie. "Theophrastus on the Eternity of the World." *Transactions and Proceedings of the American Philological Association* 71 (1940): 239–47.

McMullin, Ernan. "Conceptions of Science in the Scientific Revolution." In *Reappraisals of the Scientific Revolution*, edited by David C. Lindberg and Robert S. Westman, 27–92. Cambridge: Cambridge University Press, 1990.

————. "Explanation as Confirmation in Descartes." In *Blackwell Companion to Descartes*, edited by Janet Broughton and John Peter Carriero, 84–102. London: Blackwell, 2007.

Menn, Stephen. "The Greatest Stumbling Block: Descartes' Denial of Real Qualities." In *Descartes and His Contemporaries: Meditations, Objections, and Replies*, edited by Roger Ariew and Marjorie Grene, 182–207. Chicago: Chicago University Press, 1995.

————. "On Dennis Des Chene's *Physiologia*." *Perspectives in Science* 8 (2000): 119–43.

Mercer, Christia. "The Vitality and Importance of Early Modern Aristotelianism." In *The Rise of Modern Philosophy*, edited by Tom Sorrell, 33–67. Oxford: Oxford University Press, 1993.

Methuen, Charlotte. *Kepler's Tübingen: Stimulus to a Theological Mathematics*. Aldershot: Ashgate, 1998.

Michel, Alain. "L'influence du dialogue cicéronien sur la tradition philosophique et littéraire." In *Le dialogue au temps de la renaissance*, edited by Marie-Thérèse Jones-Davies, 9–24. Paris: Touzot, 1984.

Moran, Bruce. *Distilling Knowledge: Alchemy, Chemistry, and the Scientific Revolution*. Cambridge, MA: Harvard University Press, 2005.

Morrison, Donald. "Philoponus and Simplicius on Tekmeriodic Proof." In *Method and Order in Renaissance Philosophy of Nature*, edited by Daniel Di Liscia, Eckhard Kessler, and Charlotte Methuen, 1–22. Aldershot: Ashgate, 1998.

Moss, Jean Dietz, and William A. Wallace. *Rhetoric and Dialectic in the Time of Galileo*. Washington, DC: Catholic University of America Press, 2003.

Muir, Edward. "Why Venice? Venetian Society and the Success of Early Opera." *Journal of Interdisciplinary History* 36 (2006): 331–53.

Nardi, Bruno. *Saggi sull'Aristotelismo padovano dal secolo XIV al XVI*. Florence: Le Monnier, 1958.

———. *Studi su Pietro Pomponazzi*. Florence: Le Monnier, 1965.

Nauert, Charles G., Jr., "Caius Plinius Secundus." In vol. 4 of *Catalogus translationum et commentariorum*, edited by F. Edward Cranz and Virginia Brown, 297–422. Washington, DC: Catholic University of America Press, 1980.

Newman, William R. "Art, Nature, and Experiment among Some Aristotelian Alchemists." In *Texts and Contexts in Ancient and Medieval Science*, edited by Edith Sylla and Michael McVaugh, 305–17. Leiden: Brill, 1997.

———. *Atoms and Alchemy: Chymistry and the Experimental Origins of the Scientific Revolution*. Chicago: Chicago University Press, 2006.

———. "Experimental Corpuscular Theory in Aristotelian Alchemy: From Geber to Sennert." In *Late Medieval and Early Modern Corpuscular Matter Theories*, edited by Christoph Lüthy, John E. Murdoch, and William R. Newman, 291–329. Leiden: Brill, 2001.

———. *Promethean Ambitions: Alchemy and the Quest to Perfect Nature*. Chicago: Chicago University Press, 2004.

———. *The Summa Perfectionis of Pseudo-Geber: A Critical Edition, Translation, and Study*. Leiden: Brill, 1991.

Newman, William R., and Lawrence M. Principe. "Alchemy vs. Chemistry: The Etymological Origins of a Historiographical Mistake." *Early Science and Medicine* 3 (1998): 32–65.

Nouhuys, Tabitta van. *The Age of the Two-Faced Janus: The Comets of 1577 and 1618 and the Decline of the Aristotelian World View in the Netherlands*. Leiden: Brill, 1998.

Ogilvie, Brian. *The Science of Describing: Natural History in Renaissance Europe*. Chicago: Chicago University Press, 2006.

Olivieri, Lugi. *Certezza e gerarchi del sapere: Crisi dell'idea di scientificità nell'aristotelismo del secolo xvi*. Padua: Antenore, 1983.

Olson, Richard G. *Science and Religion, 1450–1900: From Copernicus to Darwin*. Baltimore: Johns Hopkins University Press, 2004.

Osler, Margaret. "From Immanent Natures to Nature as Artifice." *The Monist* 19 (1996): 388–407.

———. "New Wine in Old Bottles: Gassendi and the Aristotelian Origin of Early Modern Physics." *Midwest Journal of Philosophy* 26 (2002): 167–84.

———. "Whose Ends? Teleology in Early Modern Natural Philosophy," *Osiris*, 2nd ser., 16 (2001): 151–68.

Palmieri, Paolo. "Science and Authority in Giacomo Zabarella." *History of Science*, 45 (2007): 404–27.

Park, Katharine. "Natural Particulars: Medical Epistemology, Practice, and the Literature of Healing." In *Natural Particulars: Nature and the Disciplines in Renaissance*

Europe, edited by Anthony Grafton and Nancy Siraisi, 347–67. Cambridge, MA: MIT Press, 1999.

Pastor, Ludwig. *The History of the Popes*. Translated by Ralph Francis Kerr. 40 vols. London: Kegan Paul, Trench, Trubner, 1923–53.

Perfetti, Stefano. "Docebo vos dubitare: Il commento inedito di Pietro Pomponazzi al *De partibus animalium* (Bologna, 1521–24)." *Documenti e studi sulla tradizione filosofica medievale* 10 (1999): 439–66.

Petersen, Peter. *Geschichte der aristotelischen Philosophie im protestantischen Deutschland*. Leipzig: Meiner, 1921.

Piaia, Gregorio. "Aristotelismo, 'heresia' e giurisdizionalismo nella polemica del Antonio Possevino contro lo Studio di Padova." *Quaderni per la storia dell'Università di Padova* (1973): 125–45.

"Pirro Ligorio." In vol. 65 of *Dizionario biografico degli Italiani*, 109–14. Rome: Instituto della Enciclopedia italiana, 1960.

Poppi, Antonino, ed. *Cremonini e Galilei inquisiti a Padova nel 1604: Nuovi documenti d'archivio*. Padua: Antenore, 1992.

———. *Introduzione all'aristotelismo padovano*. 2nd ed. Padua: Antenore, 1991.

Preus, Robert D. *The Theology of Post-Reformation Lutheranism*. St. Louis: Concordia, 1970.

Quenet, Grégory. *Les tremblements de terre aux XVIIe et XVIIIe siècles: La naissance d'un risque*. Paris: Presses Universitaires de France, 2005.

Raimondi, Francesco Paolo. "Pomponazzi's Criticism of Swineshead and the Decline of the Calculatory Tradition in Italy." *Physis*, n.s., 37 (2000): 311–58.

Ramberti, Rita. "Stoicismo e tradizione peripatetica nel *De fato* di Pietro Pomponazzi." *Dianoia* 2 (1997): 51–84.

Randall, John Herman, Jr. "The Development of Scientific Method in the School of Padua." *Journal of the History of Ideas* 1 (1940): 177–206.

———. *The School of Padua and the Emergence of Modern Science*. Padua: Antenore, 1960.

Redondi, Pietro. *Galileo eretico*. Turin: Einaudi, 1983.

Reif, Patricia. "The Textbook Tradition in Natural Philosophy." *Journal of the History of Ideas* 30 (1969): 17–32.

Renan, Ernest. *Averroès et l'averroïsme*. 2nd ed. Paris: Lévy, 1861.

Rice, Eugene F., Jr. "Humanist Aristotelianism in France: Jacques Lefèvre and his Circle." In *Humanism in France at the End of the Middle Ages and in the Early Renaissance*, edited by A. H. T. Levi, 132–49. Manchester: Manchester University Press, 1970.

Rodis-Lewis, Geneviève. *Descartes: His Life and Thought*. Translated by Jane Marie Todd. Ithaca, NY: Cornell University Press, 1995.

———. "Un élève du collège jésuite de La Flèche: René Descartes." In *René Descartes (1596–1650): Célébrations nationales du quadricentenaire de sa naissance*, 25–36. La Flèche: Prytanée national militaire, 1997.

Roux, Sophie. "Cartesian Mechanics." In *The Reception of the Galilean Science of Motion in Seventeenth-Century Europe*, edited by Carla Rita Palmerino and J. M. M. H. Thijssen, 25–66. Dordrecht: Kluwer, 2004.

———. "La philosophie naturelle d'Honoré Fabri (1607–1688)." In *Les jésuites à Lyon, XVIe–XXe siècle*, edited by Etienne Fouilloux and Bernard Hours, 75–94. Lyon: ENS, 2005.

Ruler, J. A. van. *The Crisis of Causality: Voetius and Descartes on God, Nature and Change.* Leiden: Brill, 1995.

Schabel, Chris. "Divine Foreknowledge: Auriol, Pomponazzi, and Luther on 'Scholastic Subtleties.'" In *The Medieval Heritage in Early Modern Metaphysics and Modal Theory, 1400–1700,* edited by Russell L. Friedman and Lauge O. Nielsen, 165–90. Dordrecht: Kluwer, 2003.

Scheel, Otto, ed. *Dokumente zu Luthers Entwicklung (bis 1519).* 2nd ed. Tübingen: Mohr, 1929.

Schmitt, Charles B. *Aristotle and the Renaissance.* Cambridge, MA: Harvard University Press, 1983.

———. *A Critical Survey and Bibliography of Studies on Renaissance Aristotelianism, 1958–1969.* Padua: Antenore, 1971.

———. "Aristotle as a Cuttlefish: The Origin and Development of an Image." *Studies in the Renaissance* 12 (1965): 60–72.

———. "Experience and Experiment: A Comparison of Zabarella's View with Galileo's in *De Motu.*" *Studies in the Renaissance* 16 (1969): 80–138.

———. "The Rise of the Philosophical Textbook." In *The Cambridge History of Renaissance Philosophy,* edited by Charles B. Schmitt and Quentin Skinner, 792–804. Cambridge: Cambridge University Press, 1988.

———. "Towards a History of Renaissance Philosophy." In *Aristotelismus und Renaissance,* edited by Eckard Kessler, Charles H. Lohr, and Walter Sparn, 9–16. Wiesbaden: Harrasowitz, 1988.

Sedley, David. "Is Aristotle's Teleology Anthropocentric?" *Phronesis* 36 (1991): 179–96.

Serjeantson, Richard W. "Proof and Persuasion." In *Early Modern Science,* edited by Lorraine Daston and Katharine Park, 132–75. Vol. 3 of *The Cambridge History of Science.* Cambridge: Cambridge University Press, 2006.

Shanahan, Timothy. "Teleological Reasoning in Boyle's *Final Causes.*" In *Robert Boyle Reconsidered,* edited by Michael Hunter, 177–92. Cambridge: Cambridge University Press, 1994.

Shapin, Steven. *The Scientific Revolution.* Chicago: Chicago University Press, 1996.

Shapin, Steven, and Simon Schaffer. *Leviathan and the Air-Pump: Hobbes, Boyle, and the Experimental Life.* Princeton, NJ: Princeton University Press, 1985.

Shapiro, Barbara. *Probability and Certainty in Seventeenth-Century England: A Study of the Relationships between Natural Science, Religion, History, Law, and Literature.* Princeton, NJ: Princeton University Press, 1983.

Simmons, Alison. "Sensible Ends: Latent Teleology in Descartes' Account of Sensation." *Journal of the History of Philosophy* 39 (2001): 49–75.

Smoller, Laura A. "Of Earthquakes, Hail, Frogs, and Geography: Plague and the Investigation of the Apocalypse in the Later Middle Ages." In *Last Things: Eschatology and Apocalypse in the Middle Ages,* edited by Paul Freedman and Caroline Bynum, 156–87. Philadelphia: University of Pennsylvania Press, 2000.

Snyder, Jon. *Writing the Scene of Speaking: Theories of Dialogue in the Late Italian Renaissance.* Stanford, CA: Stanford University Press, 1989.

Sorabji, Richard. "The Ancient Commentators on Aristotle." In *Aristotle Transformed: The Ancient Commentators and Their Influence,* edited by Richard Sorabji, 1–30. Ithaca, NY: Cornell University Press, 1990.

Spranzi, Marta. "Galileo's 'Dialogue on the Two Chief World Systems': Rhetoric and Dialogue." *Archives internationales d'histoire des sciences* 55 (2005): 97–114.

Spronsen, J. W. van. "The Beginning of Chemistry." In *Leiden University in the Seventeenth Century: An Exchange of Learning,* edited by T. H. Lunsingh Scheurleer and G. H. M. Posthumus Meyjes, 329–43. Leiden: Brill, 1975.

Thorndike, Lynn. *History of Magic and Experimental Science.* 8 vols. New York: Columbia University Press, 1923–58.

Vanden Broecke, Steven. *The Limits of Influence: Pico, Louvain, and the Crisis of Renaissance Astrology.* Leiden: Brill, 2003.

Verbeek, Theo. *Descartes and the Dutch: Early Reactions to Cartesian Philosophy, 1637–1650.* Carbondale: Southern Illinois University Press, 1992.

Vermij, Rienk. "Subterranean Fire: Changing Theories of the Earth during the Renaissance." *Early Science and Medicine* 3 (1998): 323–47.

Viano, Cristina. *La matière des choses: Le livre IV des Météorologiques d'Aristotle et son interprétation par Olympiodore.* Paris: Vrin, 2006.

Waddell, Mark A. "The World as It Might Be: Iconography and Probabilism in the *Mundus subterraneus* of Athanasius Kircher." *Centaurus* 48 (2006): 3–22.

Wardy, Robert. "Aristotelian Rainfall or the Lore of Averages." *Phronesis* 38 (1993): 18–30.

Werrett, Simon. "Wonders Never Cease: Descartes's *Météores* and the Rainbow Fountain." *British Journal for the History of Science* 34 (2001): 129–47.

Zambelli, Paola, ed. *"Astrologi hallucinati": Stars and the End of the World in Luther's Time.* Berlin: De Gruyter, 1986.

Index

Achaea, 63

Aeolian islands, 63

Agathias of Mirena, 68–69

Agricola, Georg, 68–69, 82, 90–92

agriculture, 5, 46

Agrippa, Camillo, 66

Ailly, Pierre d', 3

Albertus Magnus, 3, 18, 58, 65, 68, 91, 95; on causation, 44, 127, 130, 141; on chymistry, 99, 103; on matter of meteorological phenomena, 26–27, 87–88; on Milky Way, 11; on universal flood, 71, 121

alchemy, 15, 26. *See also* chymistry

Alexander of Aphrodisias, 18, 31, 75

Alfonso II d'Este, 61–62, 67, 70, 74–75, 77–78

almanacs, 11–12

alum, 83, 85, 87, 89, 92, 95

analogy, 26, 81–82, 85–87, 117, 125; in Descartes, 135–36, 139; between firearms and meteorological phenomena, 90, 94, 97, 102, 104

anatomy, 15, 143

angels, 47, 73, 101

anthropocentrism, 39, 43, 45, 56

antiperistasis, 8, 89, 93, 99, 111

antiquarianism, 2, 60–61, 77

apocalypse, 53, 59–60

Aratus, 12

Ariew, Roger, 15

Aristophanes, 16

Aristotelianism, 98, 108, 148, 151–55; attacks on, 1, 3, 20, 149, 151; and Descartes, 131, 144–47; and eclecticism, 81, 153; and faith, 118–19; historiography of, 15–16; and humanism, 17–18; reformations of, 104–5, 115, 146; and translations, 65–66

Aristotle, 11, 15, 33, 35–37, 52, 82–85, 124; *Analytics*, 21, 25, 32, 49, 74, 115; and causation, 2–4, 81; *De anima*, 4–5; *De caelo*, 4, 17, 65, 148; *De generatione animalium*, 42; *De generatione et corruptione*, 4, 41, 46, 48, 139, 148; *De partibus animalium*, 17, 44; and earthquakes, 6, 62–64, 68–69, 75, 90–91; and epistemology, 22, 31, 109–10; on eternity of universe, 72, 119–22; *Metaphysics*, 5, 39, 49, 153; *Meteorology*, 2, 4–8, 14, 20, 24–27, 32, 35–36, 39, 42–45, 62–63, 65, 72, 83–84, 89, 93, 96, 105–8, 114, 116, 121, 129, 134–35, 144–46, 148, 151, 153–54; *Nicomachean Ethics*, 24, 32, 34, 65; *Parva naturalia*, 4; *Physics*, 4–5, 16–17, 39, 48, 148, 152–54; *Politics*, 65; *Problemata*, 65, 83–84, 92, 95; *Rhetorics*, 25; and teleology, 39–43; *Topics*, 24, 30

artificial bodies of water, 31–32, 85

artificial earthquakes, 68–69, 78–79

artificial interventions, 26, 103–4, 116

artificial models for nature, 79, 85–86, 96–99, 106, 116

artillery, 74, 79, 90, 104

astrology, 11, 13, 19, 33, 37, 48, 71; attacks on, 67, 69, 71, 73–76, 91

astrometeorology, 11, 14, 23

astronomy, 15, 24, 29, 54, 117

atomism, 56, 100, 107, 118, 126, 138, 140, 142, 146. *See also* corpuscularism; Democritus; Epicureans; Epicurus; matter theory

Augustinian order, 52, 56

Augustinian tradition, 28, 56

Avempace, 17, 29–30

Averroes, 17–18, 22, 28, 30, 44, 75, 130; on chance, 46, 49–50

Avicenna, 17, 49, 152; on spontaneous generation, 71–72; on stones, 86–87

Bacci, Andrea, 19, 92
Bacon, Francis, 1, 3, 38, 148, 151
Baiae, 85
balneology, 2, 82, 88–90, 92, 152
Barker, Peter, 34
Bayle, Pierre, 78
Beeckman, Isaac, 133
Bellarmino, Roberto, 119
Belluti, Bonaventura, 36–37, 102–4
Bérigard, Claude, 37
Bible, the, 16, 30, 68, 71, 73, 118, 121–22
Biringuccio, Vannoccio, 82, 90
Bitaud, Jean, 119
bitumen: in the earth, 36, 81–82, 85–87, 91–92, 95, 103; in exhalations, 81, 84, 86, 93–99, 102, 106, 136; in springs, 89–90, 92
Boccadiferro, Lodovico, 23, 34–35, 51
Bodin, Jean, 57, 101
Borro, Girolamo, 3, 66
Bos, Erik-Jan, 141
Boyle, Robert, 23, 38, 109–10, 150–51
Brahe, Tycho, 117
Breventano, Stefano, 66
Brussels, 80, 104, 132, 134, 149
Buoni, Giacomo, 66–70, 73, 76–78
Buridan, John, 3, 27
Burley, Walter, 3

Cabeo, Niccolò, 23, 105, 127, 141, 153; on chymistry, 106–7, 112–14, 123, 146; on corpuscularism, 106–7, 114, 123–25, 135, 147; on experience and experiments, 35–36, 110, 114–18, 120, 122–23; on faith, 118–23; on metaphysics, 108–9, 124, 128, 132; on substantial forms, 110–11, 124, 132, 150, 154
campi flegrei, 19, 94
Cardano, Girolamo, 3, 68, 95
Cassirer, Ernst, 22
catastrophes. See disasters
Catholic Church, 16, 29, 52, 67, 100–101; and Cabeo, 118–20, 123; and Descartes, 134, 138
causation, 5, 11, 13, 16, 25, 42, 47, 54, 67, 150; and Aristotle, 2–4, 81; and chance, 49–50; and Descartes, 125, 132–34; efficient, 2, 4, 14, 27–28, 38, 57, 74, 95, 111, 123, 141; and explanation, 4; final, 2, 27, 38–41, 43–44, 51, 56–57, 111, 113, 123, 127, 149–50; formal, 2, 4, 27, 111, 123, 149; material, 2, 4, 27–28, 38, 57, 78, 81, 83, 86,

93–95, 100, 111, 114, 123, 138, 140, 149–50; multiple, 19, 30, 66–67, 130; natural, 30, 57–58, 67, 69–71, 78; secondary, 55, 132
caverns, 71, 76–77, 94
celestial bodies, 11–12, 14, 34, 41, 53, 72, 150; as cause of weather, 48–49, 91, 111. See also sun
celestial region, 6, 18, 25, 152–153
Cesalpino, Andrea, 40, 44
chance, 41, 46, 48–50, 56, 58
chasms, 13, 96–97
Chiaramonti, Scipione, 153
Chifflet, Jean-Jacques, 81, 104
chymistry, 2, 19, 37, 81–82, 137, 149, 152; Aristotelian, 98–104; and Cabeo, 106–7, 112–15, 123, 146; as explanation for earthquakes, 35–36, 79, 87–88; and purple rain, 80–81; and testing, 99–102, 104, 106, 112, 115, 117, 123, 136, 142–43. See also Albertus Magnus; Cabeo, Niccolò; experiments; Libavius, Andreas; matter theory; Paracelsians; Paracelsus; Sennert, Daniel
Clave, Étienne de, 119
climate, 16
climate zones, 19, 151. See also torrid zone
clouds, 46, 102, 120, 135; and earthquakes, 63–64; formation of, 106, 125; place of, 8–9, 35; and thunder, 85, 88, 93–94, 98–99, 102, 130, 136–37
Coimbrans, 27, 108, 118, 120, 126
Columbus, Christopher, 35, 152
combination, 6–7, 9, 99, 113. See also corpuscularism; matter theory
comets, 6, 11, 13, 25, 153; as signs, 51–58, 64, 78. See also meteorological phenomena
commentary tradition, 3–4, 17–18, 20, 64, 123, 154; criticism of, 108–10; medieval, 3, 18, 44, 65
conjecture, 33–34, 76, 81, 130, 151–52
Conty, Evrart de, 65
Copernican cosmology, 100–101, 107, 153
corpuscularism, 2, 7, 20, 38, 99, 111, 117, 118, 152, 154; and Aristotle, 6–7, 83; and Cabeo, 106–7, 114, 123–25, 135, 147; and Descartes, 127–30, 134–36, 138–39; and Seneca, 84–85
Cosimo I de' Medici, 61
cosmology, 141, 148, 153. See also Copernican cosmology; Ptolemaic cosmology; Tychonic cosmology
Council of Trent, 67, 69
courtly life, 2, 11, 64–66, 71, 107

crafts, 16, 74
creation, of the universe, 43, 118–19, 121–23
Cremonini, Cesare, 3, 108, 119–20, 152
cyclones, 11, 96. *See also* hurricanes; typhoons

d'Alembert, Jean le Rond, 38
Dear, Peter, 115
Debus, Allen G., 81
deduction, 34, 128, 144
della Porta, Giovan Battista, 3, 95–99, 104
Democritus, 7, 41
demons, 30, 47, 57, 67, 77, 101
demonstration: categories of, 24–25, 32, 49; and
    Descartes, 130, 144–45; physical, 121–23; and
    regressus, 22–23; standards of, 33–34, 36, 109
Descartes, René, 1, 6, 38, 124, 142, 150, 152; *La
    Dioptrique*, 128–29, 138–39; *Discours*, 128–29,
    131, 137, 144, 146; *La Géométrie*, 128; *Letter to
    Dinet*, 143; *Letter to Regius*, 143; *Les Météores*,
    2–3, 100, 125–31, 133–38, 140–41, 143, 145–47,
    149, 155; *Le Monde*, 128, 131, 141; *Passions of the
    Soul*, 133; *Principia*, 2, 129, 143–45, 149; on
    wonder, 3, 127, 132–34, 143
Des Chene, Dennis, 15, 38–39
determinism, material, 41, 47–50, 58
diakrisis, 7, 99, 106
dialectic, 23–25, 36–37
dialogues, 64–67, 70, 76
disasters, 2, 19–20, 33, 52–53, 68; purpose of, 39,
    46, 58–61. *See also* earthquakes; floods
distillation, 81–82, 87, 89, 92, 95, 101, 103, 135
divination, 51, 67
divine knowledge, 29, 39, 50, 55, 58, 60, 67, 70
divine power, 29, 39–40, 43, 55, 67, 69–70, 75, 145;
    and angels, 73; and miracles, 120–21; and order
    of universe, 45–47, 53, 59, 76–77; and role in
    natural philosophy, 132
Dominican order, 69, 77
Dondi, Iacopo, 89
du Chesne, Joseph, 81
du Hamel, Jean Baptiste, 149–50
Duhem, Pierre, 34

earthquakes, 60–79, 81, 83, 94–96; Aristotle's
    views of, 6, 11, 14, 26, 62–64, 68–69, 75, 90–91;
    at Bologna, 76; causes of, 19, 62–64, 67–74,
    76–80, 87–88, 90–91, 94–96; at Constanti-
    nople, 77; at Ferrara, 35, 36, 60–79, 149, 152;

teleology of, 45–46, 50–53, 55–56, 75; Thales'
    views of, 26
eclipses, 49, 53, 58, 63–64
elements, Aristotelian, 6–10, 17, 37, 65, 83;
    Cartesian, 145; and form, 25, 27, 42–43, 45, 100,
    136; Hermetic, 142; rejection of, 109, 114, 119,
    135; transformation of, 41–42, 44, 72, 113
Empedocles, 41, 85
engineering, 19, 90
Epicureans, 54, 56, 140
Epicurus, 19, 130, 138
epistemology, 21–22, 34, 37, 60, 123, 149; and
    Descartes, 133, 145; and Pomponazzi, 31–32, 35.
    See also demonstration; probabilism; saving
    the appearances; signs
ethics, 32, 47, 56, 58–59, 154
Etruscans, 51
Euboea, 63
exhalations: and Descartes, 129, 134–35, 137–39,
    146; dual, 2, 13, 16, 20, 23, 27, 35, 44, 54, 64,
    81–82, 85–86, 96, 104, 110, 112, 114, 149–50,
    153–54; hot and dry (terrestrial), 6–7, 62,
    81–84, 88, 90, 93–94, 98–99, 100–102, 111;
    and substantial form, 10–11, 101–4, 136;
    subterranean, 11, 26, 35, 51, 62–64, 68–71,
    73–74, 77, 83, 89–91, 93–94, 96; vapor, 6–7, 88,
    91, 103, 111
experience, 1–2, 15–16, 20, 26, 34–36, 37, 56, 69,
    81, 104–5, 132, 151–54; and balneology, 89–90;
    and Cabeo, 35–36, 109–10, 114–18, 122–24; and
    Descartes, 128–29, 132, 136–37, 144; and
    Pomponazzi, 30–31, 46, 152; and *regressus*,
    22–23, 35. *See also* chymistry; experiments;
    observation
experiments, 2, 35, 81, 95, 148, 154; and Cabeo,
    105–6, 112, 115–17, 123; and Descartes, 137;
    and Mastri and Belluti, 103–4
explanation. *See* causation
exploration, 19, 32, 35, 151–52

falsafa, 17
fatty moisture, 8, 82, 84, 93, 96, 98, 100, 150; and
    Avicenna, 86–88; and Descartes, 137, 146;
    and Renaissance mineralogy, 90–92
Ferrara, 35, 107; and earthquakes, 19, 36, 50,
    60–61, 64, 67–77, 152
Finé, Oronce, 9
Fiornovelli, Giovanni Maria, 69

fire, 7, 83, 86, 88, 90–91, 95; subterranean,
     62, 68, 83, 85, 91–92, 95. *See also* elements,
     Aristotelian; exhalations
firearms, 82, 90, 98, 102
floods, 12, 19, 34, 39, 46, 72, 88, 120–21. *See also*
     universal flood
fog, 11, 16
Fonseca, Pedro da, 108, 118
forecasting. *See* prediction
fossils, 62, 83. *See also* thunderstones
Franciscan order, 68–69, 102
Froidmont, Libert, 100–104, 117, 136–42, 144,
     146, 153; attacks on Descartes by, 100, 126–27,
     137–41; attacks on Paracelsus by, 101; and
     lightning, 102
Frytsche, Marcus, 3, 5, 9–10, 36, 55–56, 78

Gaetano of Thiene, 3, 31, 44
Galen, 75
Galesi, Agostino, 50–51, 73, 76
Galilei, Galileo, 1, 103, 107–8, 141, 148, 152–53
Garber, Daniel, 14, 129, 141
Garcaeus, Johannes, 40, 55–56
Gassendi, Pierre, 19, 38, 81
Gerhard, Johann, 57
Gilbert, William, 3, 107
Gilson, Étienne, 125
Glanvill, Joseph, 151
Goldstein, Bernard, 34
Gonzaga, Ferdinando, 107, 117
Gozze, Nicolò Vito de, 66
Guerlac, Henry, 14
gunpowder, 79, 94, 96, 98, 102

Hacking, Ian, 26
halos, 13, 48
Hartlib, Samuel, 141
Hastings, Battle of, 52
health, 46, 56, 61
Hellespont, 63
Heracleia, 26, 63
heresy, 44, 76, 110, 119, 131. *See also* impiety
Hermetic philosophy, 137, 141–43
historiography, of meteorology, 14–15
history, 70, 73, 75–77, 79, 122, 152
humanism, 1, 5, 18, 22, 37, 53, 66, 96
hurricanes, 7–8, 11, 13, 63, 96. *See also* cyclones;
     typhoons

hypothetical method: among Aristotelians,
     22–23, 37, 47, 119; and Descartes, 125, 128–30,
     144–47

impiety, 118–19, 121. *See also* heresy
induction, 23, 115, 152
inflammability, 7, 84, 87–88, 90, 96–97, 99, 137
instrumentalism, 30. *See also* saving the
     appearances
interpretation, textual, 17–18

Jansen, Cornelius, 101
Jesuits, 37, 108, 115, 118–20, 141. *See also* Cabeo,
     Niccolò; Coimbrans; Fonseca, Pedro da;
     Kircher, Athanasius; Suárez, Francisco;
     Toletus, Francisco
Jews, expulsion of, 61, 75

Kepler, Johannes, 58, 101, 152
Kessler, Eckhard, 115
Kircher, Athanasius, 37, 110

latitudinarianism, 37
Leibniz, Gottfried Wilhelm, 38
Leijenhorst, Cees, 15
Le Vilain, Mahieu, 65
Libavius, Andreas, 99, 102
Liebler, Georg, 40, 57
lightning, 29, 57, 81, 85–88, 90, 93, 130; Aristotle's
     theory of, 7–8; and chymistry, 79, 82–84, 100,
     102, 117; della Porta's theory of, 96, 98–99;
     Descartes' theory of, 130, 135–37; teleology
     of, 46, 53, 56, 59. *See also* meteorological
     phenomena
Ligorio, Piero, 74–78
Lipara, 63
logic, 1, 14, 17, 28, 36, 102, 109, 115; in universities,
     4, 21, 23
Longiano, Sebastiano Fausto da, 66
Lucretius, 19
Luther, Martin, 52–53, 56
Lutherans, 40, 44, 51, 54, 56–57, 59, 132

Maggio, Lucio, 66, 70–73, 76–78
Manardi, Giovanni, 69
marvels. *See* meteorological phenomena;
     miracles; wonder
Mastri, Bartolomeo, 36–37, 102–4

mathematics, 32–34, 49, 107, 110, 115, 117, 128–29
matter theory, 26–28, 33–34, 100, 135, 138–39, 145; Cabeo's, 106–7, 112–14, 123. *See also* atomism; chymistry; combination; corpuscularism; substantial forms
Mayow, John, 14
mechanical philosophy, 38, 100, 127, 137, 139–40, 146–8, 150
medicine, 4, 21, 33, 40, 43, 50, 57, 67, 92, 109, 144; Hermetic, 141, 143; and scientific revolution, 15
Melanchthon, Philipp, 16, 40, 53–54, 56–58, 92
Mercurius Cosmopolita, 137–38, 141–43
mercury, 103, 109, 112, 142
Mersenne, Marin, 110, 126
metabasis, 24
metaphysics, 1, 17, 32, 102, 140–41, 146, 153; rejection of, 106–12, 115, 118, 122–24, 128, 132
meteorological phenomena, 1, 3, 15, 21, 32, 36, 40–41, 45, 51, 53, 57, 70, 79, 86, 94, 99, 101–4, 125, 136–37, 139, 142, 145, 149–50; and Aristotle, 6–8, 23, 84, 135; and Cabeo, 111, 113, definition of, 9–11; extraordinary, 4, 48, 55–56, 80, 134; fiery, 81; geological, 6, 82, 100; hydrological, 6; irregularity of, 25–27, 29, 33–34, 38, 42–44, 88; optical, 10, 48, 129–30; and Pomponazzi, 45–46; wondrous, 19, 30, 51, 65, 86, 120, 127, 132–34
meteors, 11, 65, 106
Meurer, Christoph, 56
Meurer, Wolfgang, 3, 40, 56–57, 78
Milich, Jakob, 54–55, 92
Milky Way, 6, 11, 30, 73
mineralogy, 2, 83, 86, 88, 90, 95
minerals, 86–92, 94, 98–99, 125, 136, 149
mining, 26, 82, 91
miracles, 30, 68, 70–73, 75, 78, 80, 120, 132, 134
mixtures, 6, 10–11, 28, 33, 99, 116, 138; imperfect, 2, 9–10, 20, 25–27, 38, 42, 54, 70, 100, 113, 136, 138; perfect, 10–11, 113
Mons Gibium, 92
Monte, Giambattista del, 56
mountains, 8, 34, 94, 122
Mylius, Johannes, 117

naphtha, 86–87, 92
natural history, 15
necessity, 27, 41–42, 48–51, 63. *See also* determinism, material

Neoplatonism, 17, 119
Newman, William R., 15
Nifo, Agostino, 3, 13, 21, 23, 68, 72, 93–94; on hypothetical method, 23, 33–34, 145; on torrid zone, 151–52
niter, 36, 80, 82, 85, 92, 100, 102, 142

observation, 2, 12, 22, 33, 35–36, 61, 64, 81–82, 99, 103, 131, 145, 152, 154; and Aristotle, 25–26, 63; astronomical, 117; and balneology, 89–90; and Cabeo, 108–10, 112, 115–16, 121, 123; and comets, 54; and Descartes, 128, 130, 136, 145; and earthquakes, 36, 63–64, 70–71, 152
Olympiodorus, 42
optics, 24
Oresme, Nicole, 65
Osler, Margaret, 15

Paleotti, Gabriele, 50, 71
Papal States, 61
Paracelsians, 81, 103
Paracelsus, 26, 82, 101, 104, 109, 112, 114
Paris, 44, 119
Paul IV (pope), 74
Petrarca, Francesco, 1
philology, 18, 109
Philoponus, John, 11
Piccolomini, Francesco, 21, 27
Pico della Mirandola, Giovanni, 69
Pius V (pope), 61, 69, 74, 77–78
Plato, 17, 19, 27–28, 41, 45, 53, 65
Plempius, Vopiscus Fortunatus, 81, 126, 138
Pliny the Elder, 19, 54–55, 57, 82, 86–88, 92, 95–96
Pliny the Younger, 19
pneuma, 7, 83–84
Poinsot, John, 111, 127, 141
politics, 2, 60–61, 64, 70, 74–75, 149
Pomponazzi, Pietro, 3, 21, 23, 58–59, 152; and epistemology, 23, 30–33, 35–36; and ethics, 58–59, 132–33; and teleology, 44–51, 154
Pontano, Giovanni, 53–54
Porphyry, 77
portents, 40, 51–53, 55, 58, 73, 78, 80
Porzio, Simone, 5, 19, 28, 94–95
potency and act, 16–17, 27, 110, 148
Po Valley, 61, 76–78
prediction, 5, 11, 13–14, 33, 69, 76, 95. *See also* signs

probabilism, 24–27, 36–37, 52, 60, 63, 74, 89, 103; and Cabeo, 109, 121–22, 125; and dialogues, 66
prognostication. *See* prediction; signs
providence, 16, 40, 43, 51, 54, 65, 75
Pseudo-Dionysius the Areopagita, 46
Pseudo-Geber, 98, 106–7, 113
psychology, 5, 100, 114, 117–18, 126, 128, 138
Ptolemaic cosmology, 153
Ptolemy, Claudius, 11
pyrotechnics, 90, 98, 104

radical moisture, 57, 87
rain, 11, 39, 122, 125, 132; bloody (purple), 57, 80–81, 104, 120, 132, 134, 149
rainbows, 13, 48–49, 53, 55, 116, 125, 143; and Descartes, 128–29, 134. *See also* meteorological phenomena
Ramberti, Rita, 44
Randall, John Herman, Jr., 22
Rangoni, Tommaso, 71–72
Rao, Cesare, 66
real qualities, 126, 130–31, 137–40, 143
reflection, 9–10, 101, 140
refraction, 10, 48–49, 95, 98, 128
Regius, Henricus, 131, 144
regressus, 22, 35–36, 129, 144
religion, 2, 16, 39–40, 50, 60, 70, 101, 119, 149, 154. *See also* Catholic Church; Lutherans; theology
Reneri, Henricus, 146
rhetoric, 22, 24, 31–32, 36, 76
rivers, 6, 13, 31, 65, 83, 122. *See also* floods
Rohault, Jacques, 150
Rome, 46, 61, 70, 75, 87, 101
Romei, Annibale, 73
Rubio, Antonio, 126
Russiliano, Tiberio, 34, 72, 119, 121

Sagri, Nicolò, 66
Saint Elmo's fire, 96, 130
sal niter, 90, 95, 98
Sardi, Alessandro, 67–69, 75–78
saving the appearances, 29–30, 33–35
Savonarola, Michele, 89, 91
Scaliger, Julius Caesar, 3, 131–32, 134
Schegk, Jacob, 3, 27, 34, 40, 57
Schmitt, Charles B., 16

scholasticism, 1, 17, 22, 43, 54, 107, 109, 149, 155; and Descartes, 126, 131, 136
scientific revolution, 15, 147
Scotism, 69, 102–3
sea, 122; saltiness of, 73, 116
Seneca, 19, 51, 58, 82, 84–86, 88; on ethics, 45–47
Sennert, Daniel, 57, 104, 106–7, 117, 132, 136–37, 153; and *reductio in pristinum statum*, 99–102
Serjeantson, Richard, 24
shooting stars, 7–8, 11, 13, 25–26, 101. *See also* meteorological phenomena
Sicily, 63
siege mines, 19, 78, 90, 96, 104
signs, 12; demonstrative, 22–23, 25–26, 34, 37, 62–63; divine, 40, 51, 53–56; predictive, 11–13, 56, 63–64, 73, 95
smoke, 7, 84, 87, 94, 149. *See also* exhalations
snow, 7, 11, 13, 71, 136
snowflakes, 134
Socrates, 16
Sorbonne, 119
soul. *See* psychology
Speroni, Sperone, 66
Spino, Francesco, 5
Spinoza, Benedictus, 38
spontaneous generation, 49, 72, 114, 120
springs, 13, 31–32; hot, 6, 62, 64, 83, 85–85, 89–92. *See also* balneology
Stoics, 16, 55, 58, 84, 110, 154. *See also* Seneca
Strato of Lampsacus, 91
Suárez, Francisco, 108, 118
sublunary region, 5, 11, 14, 35, 82–83, 100, 152–53; change in, 5–6, 23–26, 41–42, 48, 130; matter of, 10, 27, 37, 93, 96–97; strata in, 7–9
substantial forms, 10, 27, 39, 43, 99, 125, 150, 152; and Cabeo, 110–11, 113–14, 124, 132, 150; and Descartes, 125, 127–28, 130–32, 135–36, 138–39, 141, 143–44, 147, 150; and laboratory tests, 99, 101–4
sulfur, in the earth, 36, 82–83, 87, 91–92, 94–96; in exhalations, 80, 82, 84–86, 88, 93, 95–96, 98–100, 102–3, 105, 136–37, 142, 146, 150; and Paracelsus, 109, 112; in springs, 85, 89–90
sun, 26, 63, 68, 97, 100, 134–35, 139; mock, 13, 48–49; motion of, 7, 14, 41–42, 49, 62; power of, 14, 44, 71, 99, 103, 111, 140; rays of, 9, 13, 85, 89, 91, 98, 130. *See also* celestial bodies

sunkrisis, 6, 7, 99, 106. *See also* combination
supernatural phenomena, 61, 71, 78, 120–21, 123
syllogism, 1, 21–22, 37, 148, 154

Tasso, Torquato, 75
Taurellus, Nicolaus, 40–41, 44, 57
teleology, 38–41, 43, 45, 48–51, 55–57, 113, 149–50.
    See also causation
Telesio, Bernadino, 3
textbooks, 20, 23, 28–29, 53, 126, 150
Themo Judaei, 18, 116
Theodosius, 77
theology, 28, 39, 61, 72, 75, 80, 100, 118, 126, 144,
    151; and Buoni, 66–70; Lutheran, 40, 43–44, 52,
    54–55, 57, 59; and natural philosophy, 4, 16, 37,
    119, 121, 123; and Pomponazzi, 45–48, 50, 58–59
Theophrastus of Eresus, 12–13, 91, 130
Thomas Aquinas, 3, 18, 28, 31, 45, 49–50, 65, 69,
    77; and Jesuits, 119; on nature of substance of
    the orbs, 153
Thomism, 24, 29, 67, 69, 119, 151
thunder, 12–13, 29, 53, 56, 59, 87, 133, 137; causes
    of, 81–82, 93–94, 98, 100, 102, 130, 136–37, 142;
    and demons, 57, 101; and earthquakes, 96
thunderstones, 87, 137
tidal waves, 52, 88
Titelmans, Frans, 28–29, 69–70
Toledo, Pietro, 94
Toletus, Francisco, 108, 118, 126
torrid zone, 35, 151
translations, 4, 17; Arabic, 4, 17; French, 65, 70;
    Greek, 4, 17; Latin, 4, 17; vernacular, 65
Trevisi, Antonio, 19
Tychonic cosmology, 100, 153
typhoons, 32

uncertainty, 1, 23–24, 28, 37, 39, 66, 74, 150; and
    Sardi, 76, 78. *See also* probabilism
unctuous moisture. *See* fatty moisture
universal flood, 16, 19, 34, 53, 71–73, 121, 132

universities, 2, 11, 13, 18, 149; curricula of, 3–5, 12,
    15, 142; faculties of arts at, 4; Italian, 21, 43;
    Lutheran, 3, 16, 40, 53, 92
University of Bologna, 34, 50–51, 73, 119
University of Leiden, 143
University of Leipzig, 40, 53, 56–57
University of Louvain, 100
University of Padua, 21, 27, 35, 37, 44, 102, 108,
    119–20
University of Pisa, 5, 28, 37, 44, 65–66, 94
University of Tübingen, 40, 57–58
University of Utrecht, 131, 144
University of Wittenberg, 40, 53–55, 57, 99

Verbeek, Theo, 144, 146
vernacular writings, 65–66
Vesalius, Andreas, 56
Vesuvius, 36, 103
Vettori, Pier, 5
Vieri, Francesco de', 3, 65
Villon, Antoine de, 119
Vimercati, Francesco, 27, 35, 44, 68, 94
viscous moisture. *See* fatty moisture
Voetius, Gisbertus, 131, 144–46
volcanoes, 19, 36, 82, 86, 90–91, 103. *See also*
    Vesuvius

Wendelen, Govaart, 80–81
Wildenberg, Hieronymus, 10, 12–13
wind, 7, 11, 13, 32, 52, 75, 81, 100–101, 111, 125, 135;
    and pneuma, 84; purpose of, 45–46; sub-
    terranean, 11, 26, 35–36, 60, 62–63, 67–69, 71,
    83, 87–88, 90, 94. *See also* exhalations;
    meteorological phenomena
wonder, 3, 19, 47, 52, 65, 127, 143, 154; and
    Descartes, 3, 132–34, 154

Zabarella, Giacomo, 3, 21, 23, 28, 35–36
Zuccolo, Gregorio, 74, 78–79, 90
Zuccolo, Vitale, 66